1

" Freestyle "

I R Physics
⌄

Aaron W. Wemple

win

THE RACE
FOR
WARP DYNAMICS
RELATIVE TO WARP SPEED
AND CLEANEST ENERGY POSSIBLE

I R Physics
Published by Clean Law Publishing

International Standard Book Number: 978-0-9851567-5-6
International Standard eBook Number: 978-0-9851567-6-3

1. Physics 2. Engineering 3. Clean energy
4. Christian 5. Life Science

Printed in the United States of America
with United States Constitutional jurisprudence.

For more information:
Clean Law Publishing
www.cleanlawunion.com

Contents

Dedicated to Ace and Crush

It is impossible that God does not exist.

Only legitimate hope...

...can defy illegitimate decay.

There's been a lot talk lately about surfing the waves of gravity into the future of interstellar travel. Maybe we're getting somewhere?

The Light-to-dark Paradox Research Article

Applying ink on to paper may be the opposite or the "outside-in" of dark matter. The opposite, or outside-in action, hence understanding that is now presumed to be required to know dark matter better. Applying ink to paper is a train of thought that strengthens negatively that reinforced state of mind the more that it's applied. Maybe we need a break?

Writing may be the outside-in of dark energy. Illustratively speaking. As I type my thoughts onto the keyboard for this research article, for example, I sense compacting thoughts. Like digital eye strain on my brain from staring at a screen too long. The Light-to-dark Paradox says that we can never discover dark matter and dark energy by applying pen to paper. By outside-in compaction. So are we stuck in an inescapable rut? Should we literally escape and race to Mars?

The Dark-to-light Segment

A dark-to-light segment to this research has proven to free us to feel dark matter and dark energy better. The Light-to-dark Paradox implies that there has to be a dark-to-light segment. Even though a physical equation may never be possible for dark matter and dark energy to avoid perpetual intellectual compaction. A light-to-dark segment, or "freedom" must be needed.

So then, why proceed if dark matter and dark energy may never be mathematically documented?

Well, since the dark-to-light segment may never be shared physically on paper via mathematical equations, it could still be shared intellectually. And that's the freedom that we need to proceed. Like explaining how you first balanced and rode a bicycle. We can't go back (light-to-dark) and put into a mathematical equation how we first rode a bicycle. But we can go forward (dark-to-light) and share with others how we sensed and rode the bike.

True dark-to-light is like 100% outreach with 0% in reach. Getting as close as we can to 100% true dark-to-light and 0% light-to-dark may very well help us get closer to feeling dark matter and dark energy for interstellar travel.

From the learning to ride a bicycle analogy, the hidden energy in the universe may be presumed to be more of an inner-ear sense than it is an audio sense or a visual sense. And that's a healthy break. Hence, it should be felt and not seen or written. And this was studied in a motocross test.

Where did This Light-to-dark Paradox & Dark-to-light Segment Come From?

Since the universe is expanding at the same time that planets are orbiting one another at the same time as they are each spinning, then there are some non-perfect dynamic leverages of mass angular momentums going on. In other words, there are some imbalanced tires spinning around on cars on the freeways with all of the perfectly balanced tires. In fact, perhaps only imbalanced orbits and imbalanced spins exist in outer space?

Dynamics is the study of the motion of bodies under the action of forces. In Dynamics class in college, we calculated the distances that a motorcycle would jump given a certain weight, velocity and ramp dimensions. The jump distance could be calculated and graded for accuracy. Yet in practice on the race track, riders could easily jump more than that distance by the way that they maneuvered their bodies on the machine right before and during launch. And they could easily cut that distance by the way that they maneuvered their bodies on the machine right before and during launch. Like riding a bike, there's something felt and not seen there.

We calculated, measured and marked where they would land at 35 mph in second gear wide-open. But they laughed and went far beyond it at times and short of it at times. So, advancing physicists with freestyle motocross, gymnastics, dance, surfing, or other similar experiences may be the best practice to sense traveling beyond dark matter and dark energy.

The author of the theoretical physics book entitled *I R "Freestyle" Physics* grew up racing BMX bicycles as a hobby. And he saw how riders who used their bodies to either increase the distance from a jump to say "double" two jumps into one and decreased their distance from absorbing a jump to nullify it keeping peddles on the ground both maximized speed. Those "freestyle" racers had faster lap times.

In his adolescent years, he raced motocross and witnessed these dynamics exaggerated since the system of suspensions were added in the mix. So, all through engineering school with all of its physics, dynamics, and differential equations he looked for an answer as to why those races went faster. But, he never found the answer. Until finally he studied electrical waves and then went back to that problem. Electromagnetics is a lot of senses and very little seeing. This led to the proposed solution of the Dynamic Leverage of Angular Momentum (DLAM) Energy Theory.

The DLAM Energy Theory

The dark-to-light dynamic leverages from within a motocross jump absorbing system affects the environmental angular momentum differently than the light-to-dark dynamic leverages from within a motocross jumping system. The dark-to-light angular momentums from within a system like surfing inside of a wave tube effects the environmental angular momentum differently than the light-to-dark angular momentum going within a system like sitting on the outside of a wave. Now that may seem

15

to go without saying. But please hear me out. The energy differences felt between light-to-dark "tubular" momentum and dark-to-light "sitting still" momentum is energy in of itself. That difference is the DLAM Energy Theory. A surfer is not doing much physical work difference in those two situations. But they sure feel a balance difference. That's what needs to be exploited in the race to Mars.

We can sense that light-to-dark inertias are different than dark-to-light inertias. We can't write a mathematical equation to show others how to ride in the tube of a wave. But we can test and share those experiences.

Sitting can be written down and studied in the classroom forevermore. And that's what we're always doing. Oh sure, we test a lot of things with different sensors. But do we ever use an inner ear balance difference sensor to test for dark matter and dark energy obstacles? The DLAM energy theory is more fitting for the freestyle physics needed for interstellar travel.

Light-to-dark statics are like sitting on a surfboard on what we classically think of as the backside of the wave. We're never going to get the feel of the other side with that line of thinking.

But dark-to-light warp dynamics are like playing around on a surfboard inside the tube of a wave where we classically think of "riding" it on the front side. In particular, the "sweet spot." It's almost too fun and too effortless to believe is possible. It doesn't take much more work to balance on the "sweet spot" as it does to sit where there's no action. But the more that we start tinkering around in it, then the more that we'll be set free. The more that we'll feel the difference between those two inertias' regardless of seeing. Maybe no eyes should be involved in navigating the dark matter and dark energy quagmires while traveling to Mars?

If we're flying through outer space and try to "double" a

wave but hit a tube, then we might backflip. But if we're agile and flexible enough to use that tube as a berm instead to propel us to a more fitting place to "double" or even "triple," then we're flying with DLAM.

Does this new-thinking about literally feeling the dark-to-light side of a waves affect the environment differently than the old-thinking about seeing the light-to-dark side of the wave? Absolutely. If we use it like a play toy.

Can we intellectually quantify that difference between new-thinking and old-thinking? Or, can we just tell how "gnarly" one radical example is from another?

DLAM energy is the difference on the environment from these two inertias'. When the surface of the moon is in an expansion mode while facing the hot sun, then the initial segment of that expansion effects the entire system/environment differently than the final segment of that de-expansion from cooling. An "imbalanced tire." That different sense can be thought of as DLAM energy. And the more that we think about this and try this, then the clearer and more familiar it should become.

When the surface of earth expands while facing the hot sun, and then cools and contracts while spinning and opposite the sun, then that almost constant DLAM energy has to affect the environment around the earth to some small extent. Which should converse ripple, or suction like a "sweet spot."

When the surface of Mars expands while facing the hot sun, and then cools and contracts while spinning and opposite the sun, plus feels the DLAM energy in the environment from the earth's DLAM energy, then that almost constant accumulation of DLAM energies has to affect the environment around Mars to some extent. Maybe gravitational waves combine ripples like boat wakes? Maybe converse ripples combine into supersuctions like a vortex?

17

A Supervoid Conclusion

Everything seems to have a threshold. Perhaps balance waves have a threshold as well where they can then be seen or heard?

The more that we exploit these sensed differences, then the greater our understanding of this energy source should become.

One way to think of how the light-to-dark inertia is different than the dark-to-light inertia is like a continuity error. Like a continuity error between surfing in the impulsive "tubular" part of the wave and sitting still forever in between waves. Like "freestyle" motocross to jump and land a bike where and how we want it on the fly as opposed to where the math shows us that it should land. That's warp dynamics.

Like video gaming versus going outside and playing, a supervoid exists when the gamer is more and more self-absorbed in the program while the player outside is more and more freer. Sort of the opposite of superconductivity, and the opposite of superfluidity thresholds. A supervoid threshold may be virtually impossible to measure. But doesn't it have to be there? Is that where dark matter and dark energy is to us inside of our heads?

The supervoid threshold makes sense because that's how we classically think of dark matter. We know that dark matter doesn't interact with electromagnetic radiation (aka light) and that we only observe it because of its gravitational influence.

And maybe we can't sense dark energy because we can't sense how to get "tubular" while only testing and writing light-to-dark? Dark energy is commonly thought of by physicists as like a pressure in the vacuum of space forcing the acceleration of the universe outward.

A supervoid breakthrough, in the other universe, may be

different "freestyle" ways to feel and **exploit differences.** Analyzing differences instead of similarities might be the key to knowing and using similarities in this situation better. It might be like a smart electron traveling the circuits of internet networks to find the fastest route. And perhaps for interstellar flight and warp dynamics, we need to ride some waves, jump some waves, and absorb some waves? Obviously "freestyle!"

Could new age propulsion be just as much hyper balance, suspension and maneuverability as engine power? That's today's *I R "Freestyle" Physics: The Race for Warp Dynamics Relative to Warp Speed and the Cleanest Energy Possible.*

Introduction

I R Physics is the study and search for warp dynamics to defy the classical dark matter and dark energy quagmires in a race for interstellar travel and the cleanest energy possible.

Transformation is the process of changing form and/or appearance. Transfiguration is the final change in state of formation and/or appearance. With I R Physics, that universal transfiguration may very well be on the horizon.

I R Physics introduces several new concepts, several new theories and several new applications based on the study of systematic transformation with the potential for transfiguration.

The I R Physics books itself is a good transformational and/or transfigurational energy source. A good transformational and/or transfiguration energy source itself that naturally produces perpetually positive energy growth. These are energy producing systems by means of competitively complimentary, mutually inclusive Laws of attraction, duration and intensity. Systems with competitively complimentary, mutually inclusive methods, motives and means.

I R Physics is competitively complimentary to even classical physics.

I R Physics should open your mind to a whole new world of possibilities. A whole new world meant to reiterate the known, to better understand the yet unexplained, and to enlighten a way to the future of progress.

"Enlighten the people generally, and tyranny and oppressions of body and mind will vanish like evil spirits at the dawn of day."

- Thomas Jefferson

I R Physics has the potential for applications in catalyst free, pollution free clean energy producing systems, advances in science and technology, advances in our economy and society, as well as advances in our government and charity. And of course in our universi-ties, in our most valuable resources, and in our precious children.

I R Physics aims to discover DLAM Energy by: #1 – How can we measure transformation; #2 – What's the potential for transfiguration; and #3 – Therein lies I R "Freestyle" Physics.

A fresh new breath of life, if you will.

Because new meaning equals new life, then why not cut clean from the old and entertain something new?

"Breathe. Let go. And remind yourself that this very moment is the only one you know you have for sure."
- Oprah Winfrey

I R Physics at its core, introduces the dynamic leverage of angular momentum. And where dynamic leverage of angular momentum applies.

The ideal dynamic leverage of angular momentum can continuously grow perpetually positive. A production dynamics "equation" for transformational energy to solve perpetual motion, to solve "dark" energy, to solve the "God particle," to solve free positive power productions, and many other scientific quagmires currently unanswered.

Leverage, like anything, can be used for good and/or bad. Wisdom is knowing the difference.

I R Physics is a new frontier in the field of theoretical physics, life sciences, power production, technology, philosophy, economy, education, and society that must defy inevitable complacency.

I R Physics is the continuous dynamics of perpetually positive forces by the science of self-evidence with a brand new method, motive and means.

"The future of the world belongs to the youth of the world, and it is from the youth and not from the old that the fire of life will warm and enlighten the world. It is your privilege to breathe the breath of life into the dry bones of many around you."

- Tom Mann

Concepts

1. Definitions

Open system - An open system is one that can transfer energy with its environment.

Closed system – A closed system is one that cannot transfer energy with its environment.

Transfer – The essence of cooperative change.

Transformation – The process of cooperative change.

Transfiguration – Transfiguration is a complete change.

Mutually inclusive systems – Systems with common transfer ability.

Mutually exclusive systems – Systems without common transfer ability.

I R (Always plural in its meaning. I R has no singular interpretation.)

I R can be thought of as inner revolutionary, inner relevant and/or inner relational. Hence, no punctuation is viable. I R is the concept of systems that converge cooperatively. Therein is transformation. With the potential for transfiguration based with a dynamic universe.

I R Systems are those most perpetual in nature.

Dynamic leverage – The variable leverage of radial arm lengths that are in motion.

Dynamic leverage of angular momentum – The leverage of radial arm lengths that are in motion while rotating defines the dynamic leverage of angular momentum.

I.E.D. (Internal to Externally Dynamic) – A dynamic leveraged open force field pushing out.

E.I.D. (External to Internally Dynamic) – A dynamic leveraged open force field pushing in.

Kinetically dynamic systems – Systems that have multiple dynamic leveraged filaments that motion with respect to the dynamic leverage of angular momentum.

$c^2 \| m^3$ - is a symbolic analogy used to represent two or more competitively complimentary (c^2) mutually inclusive ($\|$) aspects of systems method, motive and means (m^3).

R.A.M.A.I.R. - is a symbolic analogy used to represent the mutually inclusive relevance of I.E.D. and E.I.D. A function of the cumulative degree of intensity effects from the dynamic leverage of angular momentum.

R.A.M.A.I.R. (Resulting Angular Momentum's of Accumulated Inner Revolutionaries.) This is the variable force field produced by mutually inclusive $c^2\|m^3$ I.E.D. and E.I.D. dynamic leveraged force fields.

R.A.M.A.I.R. is the culmination of I.E.D and E.I.D. In open systems **R.A.M.A.I.R.** $= c^2\|m^3$ (Which is an analogy.)

T.D. means the Transformational Degree of R.A.M.A.I.R.

T.P. is an analogy for the Transfiguration Potential.

T.Z. is the Transfiguration Zone.

Physics means to make known.

I R Physics' methods, motives and means are to make known better together.

I R Physics cooperates to converge with classical physics and all other true fields of study in order to reveal systems of the Light.

I R Physics is a holy derivative.

"There is no science in this world like physics. Nothing comes close to the precision with which physics enables you to understand the world around you. It's the laws of physics that allow us to say exactly what time the sun is going to rise. What time the eclipse is going to begin. What time the eclipse is going to end."
    - Neil deGrasse Tyson

"A powerful attraction exists, therefore, to the promotion of a study and of duties of all others engrossing the time most completely, and which is less benefited than most others by any acquaintance with science."
    - Charles Babbage

"Life is strong and fragile. It's a paradox... It's both things, like quantum physics: It's a particle and a wave at the same time."
    - Joan Jett

2. Revealing your I R light switch

Growing up on a B.M.X. Track, kids were enamored by the dynamics of motion that allowed some riders to go faster than others. And growing up through a university, students like me wanted to be able to understand those dynamics with the classical Laws of dynamics (a branch of classical physics.) Problem was, were no answers that I was looking for. But I knew they were there.

Revelations seem to find us in our most desperate situations. Knowing where the light switch is key.

"Dumbing down" in the traditional university, I was enamored by the false sense of security that labels like license, B.S., M.B.A. or P.H.D. made people feel. A sense of worth that may or may not even be relevant to real world value.

"All science is either physics or stamp collecting."
- Ernest Rutherford

Yet, people always seem to be paid according to their labels, not according to their real value. Problem was, I couldn't buy into that train of thought.

All the while that I was enamored with "growing up" and "dumbing down," I was also enamored with the word of God. So, God gave me I R Physics.

I R Physics became a healthy compromise for me to pursue. A way to fit in with the status quote, but a way to pursue my longing for unanswered understanding as well.

I didn't want it. God literally laid it out and then pushed it down upon me.

Turns out, many of us go through some of the same learning curves in our lives. We grow up believing one thing, and as it turns out, the reality of life turns out to be a whole other thing all together.

Hence, the potential for every reader to write their own revealing story. (See Chapter 3 – *The Ulterior Revelation Club*.)

Maybe finding ourselves outside of ourselves is where it's at?

Inner revelation that transforms us and/or transfigures us usually comes after a personal spiritual war within ourselves but due to incompatible external forces.

The challenge is accepting to forgive those forces that tried and/or did divide us.

Forgiving yourself for not only not living up to your own assumed potential if you were misled, but forgiving others whom you looked up to for guidance but guided you from their own experiences instead of your own individual intentions. The two may obviously be mutually exclusive.

The hidden key is to believe in that forgiveness. All good things will naturally aspire from there. And documenting your life from there makes life that much more meaningful. I R Physics can be your life story too.

Revealing the I R light switch is the acceptance to study competitively complimentary, mutually inclusive aspects of previously mutually exclusive methods, motives and means. And letting those systems (even mentally) converge with common, inner revolutionary, inner relevant and inner relational momentum. For that's where discovery's at.

"Part of the healing process is sharing with other people who care."
- Jerry Cantrell

"Writing means sharing. It's part of the human condition to want to share things – thoughts, ideas, opinions."
- Paulo Coelho

I R Physics has no-loose, only win-all potential for each and every one of us. Although it is very challenging. (As all good things must be.)

Now, anything holy (universal) can be represented

individually and/or as a group. Irrelevant to time and location.

Anyone can have their own I R experience. Without accepting Christ, minds get closed off, and will decay within self-imposed limitations. That's pre-I R.

Jesus is literally the switch that turns and/or keeps the lights on.

Systems like fronts in the weather, outer-space vortex holes and water flow are I R systems.

In one way, or in many ways, I R systems are continuously perpetual.

Systems that combine with variable degrees of intensity, location and duration can also have inherent transfiguration.

Transfigurational systems currently have no methods of explanation. Therefore, they are systems that are not able to be animated on accurate computer models. Systems that cannot be taught, explained or understood within classical Laws of physics.

Systems by nature that must be experienced.

Within the foundations of classical physics are self-evident limitations. Limitations that hinder progress.

Exploiting the mutually inclusive aspects of systems known to defy classical physics we find a new vantage point. We find I R Physics.

If not intuitive, just assume for a while that everything and everyone is relative to some degree, or too many degrees. And the only variable to better understand these systems is the transfer of energy, the medium of systems. That's the vantage for I R Physics.

In one way, or in several ways everything is related to some degree of relevance. Within that relevance an energy medium exists. The degree of relevance is transformational. So the degree of relevance can increase or decrease for a duration until potentially transfiguration. Like when rain, hail or lightning transfigures out of the atmosphere. Or like when you consider paying a bill, but don't really want to, but then you finally do.

I don't think that I R Physics can be as precisely defined as classical physics. Because it must consider classical Laws, but be intuitive at the same time. Firm but flexible. Bird's eye view plus worms eye view. Macro plus micro. Universal but atomic. Static but dynamics in motion.

In other words, infinitely variable dynamic concepts to resolve. Or better yet, to experience.

So therein lays the current conundrum. Do you have faith to try? Or fear to keep you back?

Do you want to climb mountains? Or sit back and

get passed up by those who do?

Does your voice want to shine? Or does your words rain?

Therein lays the potential for new discoveries applying I R Physics. Therein lies brand new potential.

For new meaning equal's new life.

3. The Inherent Ulterior Dynamics (I.U.D.) Effect

If the sphere below is variably dense (point B being twice as dense as point A) and spinning about the point C then does point A and point B have different angular velocities outside of typical length time's velocity, but with respect to density as well? (Assuming sphere is flexible.)

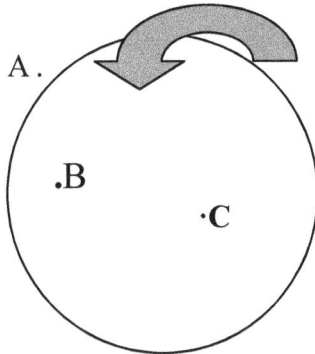

If so, then at what point does that difference cause transfiguration? At what point does the internal transforming become transfiguration?

In other words, at what point does implosion (concentration) become explosion (tear?)

Transfigurational energy is like freedom. When a new system of light inherently pulls away from all of the prior bondages that kept it tighter and tighter while spinning faster and faster past the point of implosion, an explosion, or tearing off inherently happens.

The I.U.D. Effect may be the missing link in physics. And it could have only come from God himself. As is file-stamped and recorded in the Montgomery County, IL Circuit Clerk's office, I was in jail for refusing to be an accessory to self-evident civil Rights abuse and inhumane treatment when it was discovered.

God blesses those who stand up for the truth and for those who cannot stand up for themselves, regardless.

Transformation and/or transfiguration is a result of the dynamic leverage of angular momentum.

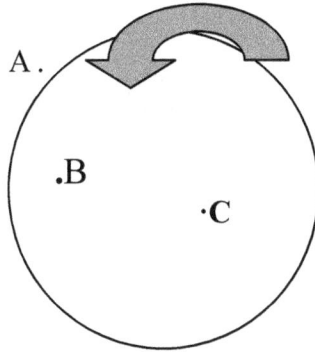

The dynamic leverages within centrifugal systems based on three complex dimensions can and will generate ulterior dynamics.

Even the invisible air has to go somewhere.

In a second example, given the following picture of an odd shaped system rotating around axis 1 it is intuitive that the resulting force at point A is greater than the force at point B. And that that force difference must go somewhere. Thus saith the Lord, a transformation of ulterior dynamics must fulfill. Which is a brief of Wemple's Law of the Ulterior Motive.

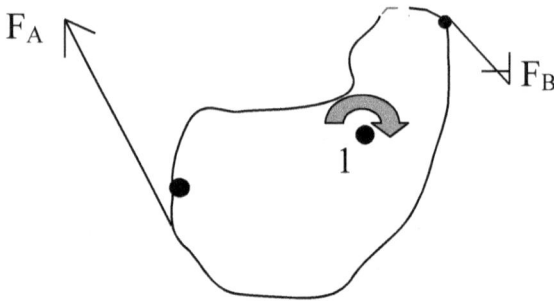

The external ulterior force produced may be a function of the systems interior radial arm length with respect mass density and the angular velocity.

The potential for division is obvious, self-evident and/or self-proving.

Likewise, if the system where malleable then internal transformation is obvious and self-evident.

The ulterior force may therefore be a function of interior radial arm lengths, variable mass density and angular velocity depending on the system.

The discoveries come when we determine at what point does mutually inclusive angular momentums become mutually exclusive. And vice versa.

If the second system were ice in water then an ulterior vortex could form.

If the system were mass in outer space then invisible atmospheric transformations could develop.

If the system were mass within mass then implosion and/or explosion could develop.

If force differences were in the air then a tornado could form.

The difference in forces of angular momentum about centrifugal systems inherently develop ulterior dynamics with a potential for transfiguration. Like lightening, hail or perpetual electricity.

Harnessing the kinetics that produce ulterior dynamics should be the pathway for advancing education, science and technology along with all of their inherent derivatives.

For those on the leading edge of progress I R Physics should help answer many of the common conundrums in theories such as time dilation (time differences with respect to mutually exclusive circumstances), no light energy, vortex holes and other currently unexplainable energy differences.

The new discoveries in I R Physics may complete the entire scope of understanding physics. And open the door to a brand new day and age full of enlightenment.

Please pray for it in the name of Jesus Christ and the world of transfiguration will begin.

I also have reports on record from prison on the discoveries of the Law of Ulterior Disordering and the Law of Ulterior Ordering. Both relating to the proprietary field of Scientific Psychiatry and Scientific Trials. (Seems Sociopathic Psychiatry enjoys benefiting off of trapping helplessly dependent victims in disorders by & for Psychotic Justice.)

Obviously, what's gamed for court is enforced to be more than two times exponentially worse on the streets.

"The possibilities that are suggested in quantum physics tell us that everything that we're looking at may not be in fact there, so the underlying nature of being is weird."

- William Shatner

The Ulterior Dynamics Club

I R means inner revolutionary, inner relevant, and inner relational.

Physics means the study to make a field, or a subject known.

I R Physics means the experience, research and means to make a field and/or subject better known by mutually accepting what's "outside of the box."

Hence, a potentially closed mind-set to overcome. It is a challenge.

"The only place opportunity cannot be found is in a closed-minded person."

- Bo Bennett

The Ulterior Dynamics Club is revolutionary. Here's how the process worked for me: I have always loved being efficient to increase productivity. I have always looked for bottle necks in systems or fields of study that I was interested in in order to try and improve them.

I researched and changed those bottle necks while remember things from my childhood that caught my attention to know what field I should look into. And I researched the Word of God almost every day.

For example, for me, one thing that I remembered catching my attention was when I was in grade school and understanding that singular judging is highly unconstitutional, immoral (if it's not with respect to God's Word), of one accord to a dictator, and just plain dumb.

Also, I learned that the military uses plural tribunals just like our Constitution orders a democratic jury to rule cases. And that just makes more sense.

Hence, "The Law Doctor" book, the Upright U.S.A. Race Family Matters Mission, New Laws, etc.

Also, I was hurt from not seeing my children for many years and wanted to pursue family court reform. Because the scales are inherently rigged.

This meant taking on the whole judicial branch of government. And trying to beat them at their own game with something better.

Hence, a mutually inclusive, competitively complimentary method, motive and means.

Another example was, I used to race bicycles and motocross, and I could understand the dynamics that made riders go faster simply by the way that they moved their bodies on the bike. Yet no university study had ever endeavored to explain how this worked.

Hence, my I R Physics work and the dynamic leverage of angular momentum discovery.

Also, I wanted to discover a way to produce free clean energy systems. This meant taking on the most advanced university minds. And trying to beat them at their own game with something better.

Hence, Forever Free Electricity.

Our competitor is best perceived as a mindset, a label or an ideology that is, or could be prohibiting field growth, preventing field progress, and/or preventing your own childhood dreams (impressions) from coming true.

In competing with roadblocks, it became "my" observation that there are systematic mind sets that prohibit the growth of not only people, but of every true field of study in systems that do not acknowledge God.

Traps already set into place by human error, wrong assumptions and/or the devil of deception alike. And these are what limit all of our potential.

At best there is no replacement for experience. At worse there is only illegitimate "replacements" for experience.

Wisdom is bridging these differences.

Assumptions are neither here nor there. Assumptions are nowhere except nowhere.

Knowing is the dare.

It's been my experience that traps are well accepted labels, a misleading mindset or false assumptions that can be deceptive by nature. A false sense of security if you will, that spreads like a virus, and limits growth.

For example: "Conservation of energy" (in physics, science and engineering), "Big Bang Theory,"

"Dark matter," "A hole in the heart" (from explaining sad emotions in psychology), and even "University."

It is "my" theory that getting stuck by willingly accepting these label traps could possibly close a mind off to new discoveries. I believe that God confirms this by telling us that those whose pride neglect to give Him credit for their achievements can suffer from strong delusions. (I think that our nation's officials suffer from this right now.)

And of course, any strong delusions by a teacher or a leader are very easily, almost compulsively accepted by any good student or any good citizen.

But this hurts very badly if those leaders are ever proven to be wrong.

That's where faith, physics and forgiveness is needed to overcome being stuck under other people's presumptions.

I like to find a way around, over and/or through those obstacles that are not proven scientifically.

(Which, even the word science comes from the Greek era relating to how they explained new knowledge that came from accepting and following Jesus Christ.)

Even written words began with men relating to God. He's always been there.

Even laws began with men relating to God.

So it only makes since, that positively perpetual power comes from men trying to relate with God.

Thus, proving by the science of self-evidence, that God must only exist. People are optional. God is constant.

And having already proven to be holy for everyone who has accepted Him, this will prove to be true for any and all of His legitimate children as well.

Anyone not following Jesus robs us of knowledge and understanding.

"Excuses" obviously rob us of knowledge and understanding.

"Protecting" "excuses," well, that's just grand theft.

Focused research plus quantity study leading to a quality outline to develop a subject anew was a good process for me beating the competitor and revolution-izing a field and subject for the story of my life. (Feel free to adjust any of this as needed for you. Just work, believe and good things will happen.)

One can change, disprove and/or even open up the mind up with better labels to breath fresh new life into a project or subject.

A fresh new breath of life means that you're still living. Never take that for granted. That's our gift.

It's been my experience to think opposite of close-minded labels works best.

For example: "Perpetual positive power" versus "Conservation of energy."

"Slow growth dynamics" versus "Big Bang theory."

"Light Science" versus "Dark Matter."

"A need for righteous acceptance" versus "A hole in the heart."

And, sometimes, "closed city" instead of "University." lol

Also, I like to think of "Republicans" as being re-labeled "Democrats" just for fun.:) They are getting too cult-ish anyway:(

And by not using names keeps it from being personal. It's not so sticky this way.

Labels on labels may or may not be a score. But

beating demons off of people that's I'm for.

I think that this breathes new life into a project that I am interested in improving with a competitively, but complimentary thought process.

I highlight, underline, and/or bold font "good" information. Keep an open mind. And then balance it all out with something more fitting. So that it has new meaning.

You already know some labels that have caught your attention. You have sensed the deadlock.

Brainstorm, write, research and document a good quantity of information over time. Especially when thoughts come into your mind.

Stay in the Word to glorify God, learn best from researching His word, and ask for His guidance.

Reference the scientific method as needed. I even think of the scientific method as being in 3-D.

If something is proved, then it's a step up. If something is disproved, well, I re-analyze everything before taking a step back down.

I reference Jesus' forgiveness formula in 3-D. When a challenge presents itself, it's either a step up or a step stagnant. Some steps can be bigger than others. (Like GIANT!) Every time you forgive it's a step up.

So, with the scientific method plus the Lord's

Law, no matter what is produced, it is right.

It's like a good virtual machine that produces only positive things.

Then, make an outline of quality topics from your own quantity of transformation information based upon that good foundation of truth and science. (Sometimes you have to "tear down" an old mind set before you can introduce something new. But that's Biblical too as far as "expose and rebuke" is concerned.)

However, I never make it person with name labels either. I tend to blame the label that my enemy is attached to. This has the potential to be a win-all.

Thus, transformational and/or transfiguration potential.

Develop an outline from a new, open perspective. Clear the table and start over. As often as it takes. Breath. Be patient. Because it does hurt your head sometimes. (Or at least it does mine.) And just feel it out.

It's like treasure hunting. But without a down side.

Actually, you're trying to beat the down side simply. Promote the good by ignoring the bad.

Remember, unstable labels can change, but some that make more sense may be needed. I try to beat the bad labels with something better. Or something more productive. This is simply like spiritual warfare.

BOOM. There it is. This is spiritual warfare. And you my friend now own transfiguration potential!

Publish regularly to protect what the Lord has given you.

Give God respect, and He will keep blessing you. Because it's easy to fall prey to strong delusions, na' sayer's, and/or to many other false senses of security. They're traps.

And those mind sets are slooow to accept anything new. And new knowledge is not easily accepted. So be patient. You may have to wrestle them once in awhile. But patients and intellectual strength are still good virtues.

Stay positive my dear friends. And purposefully driven (by God's will that is.)

And I think that it's important not to premeditate a human label or human thought onto "God's will" as well. God's will reveals it's self from the science of self-evidence, the obvious and/or the self-proving.

And when that happens, it should be a documented revelation that could have only come from the Lord. Because it is a huge new light switch turned on.

It sounds confusing, but it really is just like finding yourself and finding many other new discoveries along the way to share with the rest of the world. That's the blessing. Discovering the meaning of your life. Simple huh?

I can't tell you exactly what will come out of this work, but it will be revolutionary.

Then we can engineer a plan to put it into practice.

That's the ultimate goal.

Remember, holy revelations can be personal, universal and perpetually positive.

Your personal revelations that are also universal are probably what you're looking for.

I believe that these are revolutionary times intellectually. So the goal is to capture these new revelations in published writing for the Kingdom of Jesus.

Working for Jesus Christ realizes His will, and does it His way. It is a no-loose, win-all mind set.

And it makes the Truth that much brighter for the rest of the world to see.

So, step one to joining the Ulterior Revelation Club is to find some labels in your particular field of interest to work on.

Unless you already know of some.

Research and find some common theories and write down the science behind them. And then redevelop that science based on the old plus new philosophies.

That's applying I R Physics.

Happy surfing!:)

4. I.E.D. (Internal to Externally Dynamic) systems

I.E.D. systems may be conceptual, or even intuitive at first, but they accumulate internally by angular momentum into itself.

I.E.D. systems are inner relative to the dynamic leverage of the accumulated angular momentums of open systems. These systems are I R in nature.

In the conceptual study of I.E.D. systems within I R Physics, only the mutually inclusive (their competitively complimentary aspects) method, motive and means need to be considered. And those aspects may differ for different systems.

Classical physics has dealt with probably every possible mutually exclusive aspect of systems that there are. But the two (mutually inclusive and mutually exclusive) don't relate. They only deviate.

Therefore, let's take into account a low pressure, I.E.D. weather system first. Four degrees of reference may be needed to understand them accurately. With a relevant angular momentum of 1% "B position" within a low pressure I.E.D. (the whole system being 100%) weather front, there are obviously degrees of accumulation since that 1% can competitively compliment (or c^2) other 1%'s. And the degrees of accumulation can inherently decrease and/or increase.

And these degrees are relevant to five dimensional space, material, speed, temperature and their surroundings. Even a sixth dimension of angular momentum with respect to other angular momentums. Even a seventh dimension with respect to external IUD Effects.

These accumulations may add up to such a degree as to allow a transformation output. This is what I call the Transformational Degree (or T.D.)

As any good radar system shows, the two dimensional angular momentums are cumulative.

But the current radar systems can account for a third positioning dimension and some-what a fourth dimension of density. But not an infinitely fifth, six or seventh dimension of variable angular momentum's.

And current computer systems have limited account for transfiguration potential.

Adding a fifth, sixth or seventh dimension to centripetal concepts on today's radar systems, and it should be easy to understand the degree and difficulty describing these forces within the classical Laws of physics. As well as the difficulty corresponding reality to computer animated programs.

Classical physics and the new may not even relate.

And as of present, many of these forces have even been perpetual themselves in nature. In other words, never ending.

Therefore, how could any static Law have ever, or will ever explain them entirely? I don't believe that it's possible.

So, proportional duration, mutually inclusive relevance and/or degrees of relevance (intensity) looks to be the best way to conceptualize them. Even for their potential for transfiguration. Their maximum Transformation Degree (T.D.)

Even the power factor that may result from transformation needs to be further researched to understand in order to use the excess energy concept of the R.A.M.A.I.R. Effect.

Next, let's take a deeper look into the cosmos. Material (a variably closed system) and space (a variably open system) are competitively complimentary to the degree that closed systems don't impact each other. This T.D. of potential impact can be estimated by the quantity of dynamic material within the mutually inclusive space.

At the instances where closed systems interact, the $c^2 \| m^3$ (competitively complimentary mutually inclusive ($\|$) method, motive and means) concept ends, and classical physics takes over for a duration.

All of these systems currently have known unexplainable forces.

I R Physics may be better for understanding "chronic" conditions. While classical physics are better for understanding "acute" conditions.

It's my hypothesis that all of these unexplainable forces can be explained through the cumulative action of mutually inclusive angular momentums. The common aspect to open and closed systems. The "glue" of the two, if you will. A virtual, but known field of force. The hallmark of I R Physics as it applies to physical nature.

Angular momentum, as opposed to classical physics, must be experienced to get a true sense of it.

If no organ in the human body can decipher angular momentum accurately except the inner ear then how else can one even relate to it unless they legitimately experience it?

Given these new concepts, it seems logical to me to leave out classical Laws of physics for the time being, in order to get a better grasp on mutually inclusive dynamic systems more thoroughly.

But that's the way any new science or discovery starts out, from scratch.

5. E.I.D. (External to Internally Dynamic) systems

E.I.D. systems may be conceptual, or intuitive at first, but they are accumulated externally by angular momentums towards another system.

E.I.D. systems are inner relative to the dynamic leverage of the accumulated angular momentums of inner revolutionaries. These systems are I R in nature.

An E.I.D. system is the same type of system as the I.E.D. system when they are mutually inclusive. E.I.D. is simply the exploding of the two for reference.

Like any particular aspect of physics (light, sound, sight, temperature, pressure, gravity, etc.), a proper proportion may be a fine attribute to find in order to understand I R Physics completely. That's our window.

Given light sources, and light emitters, color itself is a very tiny attribute of the overall scale of light waves that are visible to the human eye. And thus, mutually inclusive to people so that we can study them and use those studies.

Audible sound waves have a similar fine attribute that is mutually inclusive with human beings ear/nerve/storage systems.

Gravity is on the same fine scale.

All of our human senses are within a finite scale with nature.

Understanding and applying those mutually inclusive senses are the transfigurational miracles in life, science, education, etc. For that's where healing is.

When an E.I.D. system is much greater than an I.E.D. system, then the I.E.D. system may simply be absorbed into the E.I.D. system without a meaningful effect.

Multiple systems and multiple layers also add to the difficultly of a finite definition.

Viable finite I.E.D. aspects may also very well differ from viable finite E.I.D. aspects. They're inherently transformational. And if they are too different then they are inherently possibly transfigurational.

If two audio speakers put out 100 decibels each, and one person is listening to both of them, does the listener receive 100 decibels, or 200? Could ten of these speakers increase the relevance?

How about if 100 people listened to one 100 decibel speaker. Is that 100 decibels for each person, or 1 decibel for each person?

The idea here is that the speaker and the receiver (human ear) are both open systems. And the only way that they can relate is by a dynamic ulterior experience that both share. In other words, mutually inclusive aspects.

This mutually inclusive aspect seems to be the key to open-open systems.

One or many dynamic variables seem to be the means.

Advancing science and technology is the motive.

If systems are open, then there are virtually an unlimited way of relating every person, from every angle, from every distance, etc., etc., etc. Therefore, I.E.D. and E.I.D. aspects of the sound should be analyzed with respect to the source and the receiver's mutual inclusiveness to find a common denominator. (Or the R.A.M.A.I.R. Effect.)

Can a closed stereo system ever be transformational by itself? No.

Can a closed mind ever be transformational by itself? No.

Can the sound waves that an open stereo system produces be transformational with respect to a closed mind that hears it? No.

Can the sound waves that an open stereo system produces be transformational to an open mind that accepts it? Yes.

Even transfigurational? Yes, with respect to a

mutually inclusive relative method, motive and means.

But, you may ask, if a tree falls down in the woods, and nobody is around, does it make a sound? And the answer is no. Because "sound" is a label that open minds use to share the experience of audible waves. If there is no open ability to put a label on a thing, then it has no meaning, less obviously delusional. (Which is a possibility to preserve individual pride.) It's a trick question put out there by people trying to inherently manipulate submission.

If a tree falls, sure it falls. But if you accept the bait and "think" that it could possibly make a sound, then it is self-evident, you become a little bit delusional. A little bit weaker. A little bit easier to be taken advantage of. And you now become more submissive to whoever asked you the question simply because you have bought into chasing a ghost.

The point is, don't buy into even potential tricksters. Learn for yourself and grow perpetually positive inside.

Even if you have to learn from your own mistakes. Because that is after all the most meaningful way to learn.

If I closed my ears and talked, then that's louder to the closed system (my brain) than if I opened my ears and talked. But nothing changes for the open system.

Therefore, relevance for open-open systems is proportional to the degree of energy transfer. Not the energy source or to the energy receiver.

The right spoken sound waves really seem to have an unlimited potential for transformation. Potentially greater than all other energy sources.

Words spoken thousands of years ago could really have more meaning today than ever before. And that's positive power energy dynamics in action. That's more energy than even the sun can continuously produce. (Since the sun loses energy over time.)

So, does sound have more meaningful power potential than the sun? Obviously!

Does the accumulation of human beings who only accept the truth have more power than even the sun? Yes.

Therefore, righteous speech is obviously a trans-formational and/or a transfigurational source of energy. (Though not as time sensitive compared to all other energy sources. But mutually inclusive application, duration and intensity sensitive.) And most likely defies classical Laws of physics outside of the standard time scale (i.e. open atmosphere.)

Therefore, the mutually inclusiveness of truth and the human relevance obviously defies the 2nd Law of Thermodynamics. Also known as the Laws of decay. Which state that everything "must" decay eventually.

Truth plus human relevance obviously grow, not decay. This is the First Law of Human Authority.

The First Law of Human Authority can & will be tried and proven to be true by either the ace of faith, or else by the science of self-evidence through trumps of truth (documented trial, error, reason, and why.)

Are people who accepted spoken words and their actions the only relevant variables? No.

But those who have always accepted the truth, and continue to persevere in that direction, obviously have the one and only true precedence that has ever existed.

The longevity and intensity of mindful resonance is obviously another factor. Another continuously open force field.

Thanks to our constitutional founding fathers, civil soldiers and many other great leaders, this mindful resonance blesses many of us today & tomorrow.

Does all sound defy classical physics? No. But truthful sounds do.

It's like faith compared to fear. They don't relate. They're mutually exclusive. Therefore, they're only truly known by there own.

Are the right sound waves alive irrelevant to instances, systems or circumstances? I believe that is self-evident, yes.

Prove me wrong and/or prove me right on this and we will all mindfully advance. This is a no-loose, win-all challenge.

Hence, the best communication is perpetually positive. Those good seeds presented and then accepted grow with duration and intensity, regardless of standard time scales. And regardless of negative noise.

A true, open ended, positive powered, dynamic leveraged system that no sane mind can deny.

Naturally, obviously, self-evidently, and even self-provingly it can only have one heavenly source.

Prove God wrong and/or prove God right, and "you" will only eventually grow to prove that God must only exist.

I R Physics is, therefore above all, challenging to the contrary.

6. R.A.M.A.I.R.

R.A.M.A.I.R. is the Resulting Angular Momen-tums of Angular Inner Revolutionaries. A new transforming and/or potentially transfiguring energy source.

R.A.M.A.I.R. is an attractive new force for a perpetually positive powered frontier for the broadest horizon.

Since the competitively complimentary aspect of open and open systems is angular momentum, then that's our focus. That's R.A.M.A.I.R.

And at some mutually inclusive degree, R.A.M.A.I.R. is transformational. That's the Transformational Degree (or T.D.)

This T.D. can have a variable Power Factor (P.F.) with respect to the systems themselves. Similar to the F-scale for tornadoes, the Richter scale for earthquakes, or the Scoville scale for hot sauce.

"What we usually consider as impossible are simply engineering problems... there's no law of physics pre-venting them"
- Michio Kako

Every other field of physics has a mutually inclusive, transformational aspect; light waves from source to vision, critical temperature for superconductors and electrons, gravity for space and mass, electricity for generator and motor, etc., etc. Those are open and open systems. Even if dynamic leveraged. Those already have a T.D., or Transformational Degree of relevance.

R.A.M.A.I.R. is the result of open and open systems. The main focus of I R Physics.

I define the R.A.M.A.I.R. concept as the dynamic momentous field of force where the I.E.D. force field meets the E.I.D force field. R.A.M.A.I.R., therefore, can only have a potential for transformation. In other words, transfiguration may or may not happen. (Like water going down a drain that gains a centrifugal force component and/or vapor locks. It depends on the circumstances.)

The R.A.M.A.I.R. force field is the energy transfer from the dynamic leverage of mutually inclusive systems.

R.A.M.A.I.R. is a counter intuitive concept to classical physics.

For physical systems, this R.A.M.A.I.R. potential field may be angular momentum. For electromagnetic systems, this R.A.M.A.I.R. potential field may be electromagnetic instances. For multi-substance systems (tornadoes, galaxies, etc.), the R.A.M.A.I.R. potential field may be flexibly variable per the system's competitively complimentary methods, motives and means $(c^2 \| m^3)$. Combinations of physical angular momentums and/or electromagnet instances. ($c^2 \| m^3$ is an analogy to represent competitively complimentary, mutually inclusive methods, motives and means.)

Newton's Laws of classical physics are all accurate, but end at the "cannonball" theory (used to plan outer space travel.) The cannonball theory is a partial I.E.D. system.

A complete Law would consider the R.A.M.A.I.R.

Physicists have yet to accurately define "chaos" forces, multidimensional water flow, and weather turbulence for many cases. I R Physics could get them there.

Einstein's theory of general relativity (a proposed relationship of gravity, space and time) has very minute discrepancies near the same limits. But it does not consider angular momentum potential.

"Dark matter" (a hypothetical form of matter postulated to account for the unknown forces in the universe) or "chaotic motion" theories attempt to compensate where the Laws of classical physics have not, or cannot endeavor. They are limited because they cannot accurately consider the mutually inclusive aspects of open and closed systems together.

Pseudo forces (any force assumed by an observer to maintain the validity of Newton's second Law of motion in a reference frame that is centrifugal or accelerating) are known to be non-classical as they violate Newton's third Law.

But, taking into consideration that "pseudo forces" attempt to explain open and closed system relevance, it's much like comparing material with space (i.e. apples and orange juice smell.) Pseudo forces as well are not real because they are not medium oriented between open and closed systems, nor open and open systems.

A Coriolis force attempts to explain the same closed system versus open system, but the angular momentum aspect is mutually exclusive. So it's irrelevant as far in any new study. It differs for the open system with respect to the closed system.

Therein lays the self-evident conundrum (puzzle) to solve. Concepts that need to be broadened in order to understand the mutually inclusive aspects of systems properly. Because in reality, everything can be mutually inclusive at some degree or another by the science of self-evidence.

The R.AM.A.I.R. concept is the key, therefore, to relating the angular momentum (the mutually inclusive aspects) of open-closed and open-open systems.

And the degree of the mutually inclusive I.E.D. and E.I.D. relevance is the R.A.M.A.I.R. intensity.

Everything in classical physics is precisely fine-tuned at similar mutually inclusive levels. Light, sound and our other senses are mutually inclusive to people and our atmosphere at precise levels for them to exist. Our instruments have a mutually inclusive range with their surroundings and with their observer.

R.A.M.A.I.R. is a similar concept, but only our inner ear (balance) can sense these angular momentum changes. Therefore, it is a "new" dimension as well. The R.A.M.A.I.R. dimension(s.)

A R.A.M.A.I.R instrument(s) may be needed.

Surfers, racers, gymnasts, etc. have a greater aptitude for these senses.

But also, like light instruments trying to measure sound, not all minds are open enough due to inexperience.

Therein lays another self-evident stumbling block. Most surfer types are not the scientist types. (Although there are very good exceptions.)

I propose that the research, understanding, and study of this new I R field of physics can only help solve many of the world's most meaningful problems that either constructs or deconstructs anything monumentous in life that we have yet to fully figure out.

I R Physics may even accurately improve oppressed social groups, advance clean energy technologies, and apply a better stewardship of our atmosphere and earth.

Especially critical, is it's potential to know transformational points of the earth's angular momentum with respect to coal, rock, oil, and natural gas displacement. Transfiguration is known to be planetary.

In my ongoing studies, my initial goal was to harness centripetal forces to more efficiently produce electricity from a generator, wind turbine or motor-generator set.

My motive has led to a completely different way of understanding physics as we know it today. And has as of now seemed to open the door to very real new ventures deemed possible only with this new cooperative mind set. Prayerfully, a whole new field of physics.

But, please be aware, concepts of this magnitude may be psychologically treacherous if not pursued patiently, and properly purposeful. Because they can and will hurt your head if you think on them too long at a time.

The R.A.M.A.I.R. Dynamic Force Field Theory of mutually inclusive I.E.D. & E.I.D. affects:

With proper discernment and balance, the dynamic leverage of inner revolutionary physics maintains Newtonian Laws and geometric energy conservation while infinitely varying torque energy of an entire system by the dynamic leveraged angular momentum's, radial arm lengths (even virtual within electromagnetic and other systems), momentous mass configurations, and/or electromagnetic forces and configurations, even possibly spectrum wave configurations, about multiple axis and multiple perspectives. Where the duration, intensity and/or relevance of angular momentum become the only concept of universal study.

I believe that I R Physics could help better explain and understand outer space vortex holes, star implosion and explosion, tornadoes, nuclear physics, etc., where general relativity, Newtonian Laws and other propositions either leave unanswered, or muddy up the waters.

I can visualize so many more applications of this study. From sports safety equipment to psychology. From weather to new solar panels.

On the humorous side, a surfer might be able to ride the swell of an outer space vortex hole without disintegrating.

A surfer might even be able to ride the inside of a tornado. But who else would even try?

Or, more practically, we may be able to apply motion sensors to a balance-able surfboard and send it near a giant water drain, and computer model both, the surfer and the drain water to better conceptualize the I.E.D. & E.I.D. affecting inherently R.A.M.A.I.R. effects by inclusive angular momentums.

I believe that I R Physics would lead to a better understanding of superconductors. Which could lead to perfect electronic systems with no need of batteries after start up.

Current superconductors can maintain an electrical current indefinitely because of the phonon effect. A truly efficient energy transfer system.

I R Physics may explain the currently unknown reasons for newer, non-conventional superconductors that have a high critical temperature for zero resistance. This would be a transfigured system on the market.

Superconductivity at room temperature may or

may not be possible. Classical physics is sensitive to time and differences, where I R Physics is sensitive to cooperating duration, degrees (variable intensity) and relevance. So we may soon find out.

I believe that the I R Physics of systems need their own field of study for the most efficient ease of applications and acceptance. A new field of study based on a Standard of Symbolic Represented Truths (S.S.R.T.'s) derived from the new and mutually inclusive I R Physics perspective along with classical physic.

I've discerned that I R Physics may hold the key to more accurate weather prediction, tornado destabilization, nuclear radioactive clean up, ozone clean up, clean power technologies, and non-lethal transportation, non-illegal transportation, batter laws, better family authorities, better government, advanced psychological, biological, economy, and social systems and ventures alike. An investor's paradise.

The mutually inclusive aspects (good mind sets) of good sound waves can even be considered for their transformational degrees of potentials for transfiguration for applications into practice. Even after thousands of years from when they were said. And that's a divine miracle. As all self-evidence has proven.

"Before the discovery of quantum mechanics, the framework of physics was this: If you tell me how things are now, I can then use the laws of physics to calculate, and hence predict, how things will be later."
- Brian Grenn

Could the invisible I R perspective be the
"Parallel" universe that some describe? The same
considerations, just with a fresh new perspective?

Could $2 + 2 = 4$, and 1 trillion $- 999,999,999,996$
$= 4$?

It's possible.

It just depends on the shared medium.

Let's race!?

7. The R.A.M.A.I.R. Effect

The R.A.M.A.I.R. Effect: The mutual inclusiveness of systems inherently harnessing variable degrees of relevance, variable degrees of intensity and variable duration.

     I have discerned that energy transfer will be more diligently studied, and therefore better known and used, by means of I R Physics. A new concept competitively complimentary with classical physics.

     Energy and its transfer means may be better understood itself, even self-evidently, as the dynamic kinetics (motions) from each I.E.D. and E.I.D. perspective with respect to their mutually inclusive R.A.MA.I.R. potential field of inclusion.

     A new comprehensive, more diligent look at transformation (the process of changing form or appearance) energy and/or transfiguration (the final change in form or appearance.)

     Like a tornado (transformational and potentially transfigurational R.A.M.A.I.R.) field, depending on the desired energy transfer understanding, from the larger atmosphere perspective, as well as from the ground perspective. A much broader consideration is understood and would be beneficial.

     Regarding a leaf floating down a stream, the multiple I R Physics concept considers a study from the shores vantage point (E.I.D.) and the I.E.D.

vantage point of the water turbulence as they relate to the surface were the leaf floats. Both vantage points should be geometrically considered per angular momentum in multiple dimensions for accuracy and means where they correlate. Hence, the mutually inclusive gravity, and capillary action between the leaf and the water would be better known.

A leaf floating down a stream would be a closed system atop of an open system governed by a closed but variable shoreline. Hard to document. But not so hard to experience.

Gyroscopic (an open spinning wheel governed by a closed boundary) effects are another example of an I R Physics system to further consider. An I.E.D. system within a closed system. An instance of consideration without this boundary would be similar to the R.A.MA.I.R. dynamics effect. (Though on a small scale.)

With regards to the torque that a gyroscope produces, it obviously defies classical "conservation of energy" theories. Because after initiation, it perpetu-ally produces torque with angular momentum.

Many of us have considered holding a spinning bicycle wheel in our hands by its axle. If this spinning wheel had no boundaries of restrictions (like a tor-nado or cosmos), and were free and flexible enough to break apart, piece by piece, part by part, concussion, and repercussion as each part went, the R.A.M.A.I.R. potential field, I.E.D.'s, and

E.I.D.'s should be self-evident. As well as their difficulty and willingness for understanding. It is a challenge. As all good things must be.

Even an I R nut being tightened onto a bolt has multiple vantage points and multiple frames of refer-ence. Many have heard the concept "righty tighty." Well, that's okay if the perspective is from top dead center of the nut from a top view point.

However, from my hands perspective if I'm tightening the nut with a wrench then it's "lefty tighty." Unless I'm using a ratchet (another I R device), then my hand goes right and left to tighten and/or loosen the device.

But if I've got two ratchets coupled together end-to-end, both ratchet-able (dynamic leveraged I R Physics) my desired outcome can be describe by many different means depending on needs, and 1-D, 2-D, or 3-D perspectives.

Much like the inner workings of a cosmos cluster or a tornado. (Which have dynamic kinetics instead of static leverage.)

To get a true grasp on many I R systems, a multiple dimensional perspective field should be considered, and independently studied to verify. If nothing else, because of the complexity of these fields.

Only then would the R.A.M.A.I.R. force field be considered accurate with reference to both I.E.D. and E.I.D. perspectives.

And only then could angular momentums of variable degrees be known as a new force to work with.

A universal equation that takes into account multiple perspectives, but with flexible degrees of intensity with respect to open systems may actually exist within these new concepts.

The "parallel universe" currently being considered by many physicists may actually be a view point from this same perspective since it too is an invisible dimension.

A divine point of view, like a cursed point of view, is provable and well known for social success and/or social failure. And that's the no brain-er to further research I R Physics. Even if it appears futile to some, everyone will gain something by advancing it's meaning.

I don't believe that the universe could have happened in an instance. I don't believe that there was ample means, material or motive for a closed system to explode. There must have been at least oxygen present to enable explosion. Only in open-open systems do we see transfiguration. And only then, after some duration, intensity and relevance.

Like next year's trees growing from its own, a properly proportioned season of transformational energy to produce a transfiguration seems to be needed for everything derived from closed to open systems. A competitively complimentary method, motive and means. Without God, a mutually inclusive, that's impossible.

Plant seeds can be understood as a closed system governed by the Laws of thermodynamics (decay) until they are planted. That's when they become mutually inclusive to variable degrees with an open soil system, temperature, water, etc.

Now soil is decomposed stuff. It isn't natural for decomposed stuff to give life. In fact it's contrary.

But add in the concepts of dynamic kinetics with temperature and variably circulating water and atmosphere, the pressure and depressure of the atmospheric temperature for a competitively complimentary method, motive and means, aspects that breathe life into a system, and the closed seeds become transformational with open nature and may sprout transfigurationally.

Now, when the plant grows past the soil, it faces another challenge of its own by being a part of the mutually inclusive atmosphere while being part of the soil. This begins another $c^2\|m^3$ transformational and/or transfigurational process. And the added E.I.D. mutually inclusive aspects are the light, water and nutrients (since the temperature is mostly the same for both.)

But when the plant matures and the old roots and old life die, a transfiguration may begin. Several new closed systems have formed (seeds) within the entire inclusive dynamically kinetic open-open system. In other words, something from a previous nothing.

Just like people, we could not have come from nothing. That's self-evident.

Which came first, the chicken or the egg?

Neither. God must only have come first.

Open-open systems can produce closed system, but no other mix of systems can.

And without a permanent, "initial" open God, nothing was, is or could ever be possible.

Like love, life, science, biology, the great things that last and grow, a "perspiration" or "breathing" of the universe seems to be a Holy catalyst. In, out, up, down, competitively complimentary method, motive and means are needed for good open systems to suc-ceed forever.

Closed systems are there for a footstool, if you will.

But open-open systems can develop their own self-evident new foundations. Even if it's a new, transfigured foundation.

Closed systems are optional. Open systems are not.

From any perspective, however, almost all other current universities are pretty closed minded about the important matters today. People naturally become self-serving and only self-evident. A Presti-gious Complex Disorder (P.C.D.) that all servants of a closed system falls into eventually. If, by nothing else, by pride. Complacently stuck in unassuming mind-sets. Due, I believe, to the old myths of "conser-vation of energy," "evolution," "survival of the fittest," "product of environment," "it has to be this way," and the marketing effects on society that we've all been taught, bought into, and are now indoc-trinated with, or else be considered as an outsider.

Closed labels can become a handicap to progress. They can and will rob us of knowledge and understanding.

Closed labels can and will eventually become a handicap of Biblical proportions. In fact, we should already be starting to think about of a new label for I R Physics today.

Open-open systems are the best because only they have the potential to harness new universal truths. It's known to be this way.

Closed minds loyal to closed labels or closed systems too big end up oppressed. It's known to be and this way.

And when minds are already oppressed then they're very easy to suppress. And when they're very easy to suppress then they and their children are very easy to inherently traumatize. At this point, making zombies is easy and/or easier by simply neglecting them. Especially by neglecting the "best" with excuses.

Some seeds can have no chance without an open system to mature in. Labels, buildings, complacent guidance, illegitimate "authority," and things can be a handicap to even the best minds. And sometimes with force if "need" be.

New breaths of life to complacent systems are never fully acceptable at first. But duration and intensity of the most accurately diligent minds prevail over all. As logically, only they can. Not the labels or the closed bad things. They are simply footstools to be used. As logically, only they can.

Only when they become knowingly competitive are those new mind sets accepted and then taken seriously.

However, if they can remain competitively complimentary, well, then that's a win-all, regardless.

And that my dear friends was the example that Jesus Christ left for us to live. For none of us are perfect. We all fall short. We're all pushed down. We're all held down. We sometimes push others down. And we sometimes hold others down. Even if it's something like telling them something that you thought was right but then it turns out that it was wrong.

And it's impossible for God not to exist when we grow back up after these facts through no "fault" of our own.

"An egg is a beautiful, sophisticated thing that can create even more sophisticated things, such as chickens. And we know in our heart of hearts that the universe does not travel from mush to complexity. In fact, this gut instinct is reflected in one of the most fundamental laws of physics, the second law of thermodynamics, or the law of entropy."

- David Christian

Yet hope still floats. Dare I say hope even breathes?!

8. The finish: I R Physics

Variable space pressure coupled with variably dense material may be more of a matter of inherent conditions than of measurable substances. Not so much time and mass related (though these are useful tools), but more degrees and relevant to conditions related.

"Physics is, hopefully, simple. Physicists are not."
- Edward Teller

I R Physics is the consideration of these alternative means, matters, and conditions interacting in a com-petitively complimentary new way.

And a fresh new perspective on space, time, life, science, technology, economy, society, biology, govern-mental, psychology, and the likes. Because the under-standing of physics is universal.

Much like entropy in thermodynamics, space and materials are known to be more a matter of the con-ditions that we examine them in. But we still find it useful to relate and reference them with respect to time and mass, not mutually inclusive relevance.

But decay (entropy) is more proportional to material and atmospheric conditions than to time or material. That is obvious when researchers note that atmosphere temperature has more to do with physical decay than time does.

Entropy is the study from the closed systems perspective. From an open systems perspective (soil and/or atmosphere), they actually gain something.

This is the relevance to resolve. Understanding this transfer better, and I R Physics is better understood.

Composition and decomposition alike must be better explainable using a relevance scale like R.A.M.A.I.R. Or, by attractive degrees of I.E.D. and/or E.I.D..

The time and mass standard is the status quote in classical physics. But, like living versus dead things, standards of observation obviously differ, but duration and intensity verifies.

And so should there acceptance of new competitively complimentary methods, motives and means.

Time seems to me to be a closed system. A potentially closed mind-set. A human originated tool much like any label that people use to relate to each other, and then get used to. And can therefore be a rut. (As people unfortunately become loyal to labels.)

The time scale method, motive and means may or may not apply to other conditions if different, non-proportional or outside of that box.

Could one day from the Lord's perspective really equal one thousand years from our perspective? YES!

With Newton's Laws of classical physics (formulated truths) we can design and build physically stable systems; skyscrapers, bridges, vehicles, etc. The Newtonian mechanics that provide the tools for these endeavors where the first truly revolutionary develop-ments in theoretical physics which became irrevocable Laws of science, society, economy, and technology. That can never change. But it may improve.

Examples of our faith in Newton's Laws are the studies into galaxies and clusters of galaxies that are observed to rotate. From these Laws, we can calculate the amount of matter that "must" be present in the galaxy or cluster for gravity to supply the centripetal force corresponding to the observed rotation.

Yet, the amount of matter that we can actually observe with telescopes is far less than we expect. So what gives? Where's the extra force coming from?

Therefore, it has been proposed that "dark matter" not seen with telescopes but that "must" be present to provide the needed "gravitational" forces. "Dark matter," that I believe, can be explained by the I R Physics of angular momentums produced by dynamic leverages. Even by inherent dynamic kinetics.

There is yet no other fully convincing candidate for the type or nature of this "dark matter," and so other explanations have been proposed for the inconsistency between the amount of matter observed in the galaxies and the amount needed to satisfy classical Newtonian Laws. (Apples and orange juice smell?)

In the past few decades, an apparent revolution has emerged. This development concerns mechanical systems whose behavior is described as "chaotic." How else can you be looking at apples but sense orange juice?

One of the hallmarks of Newton's Laws is their ability to predict the future behavior of a system, if we know the forces that act and the initial motion.

For example, from the initial position and velocity of a space probe that experiences known gravitational forces from the sun and the planets, we can calculate its trajectory. We see apples and know where they go.

On the other hand, consider a twig floating down a stream. We cannot currently sense where it will end up. Likewise, we cannot currently know weather patterns, waves toppling over, water draining, and the cosmos effects that harness similar attributes. Attributes that can obviously be accumulative. In other words, we know where the apple is falling but we can't figure out where the smell of orange juice is coming from.

So, what is the common denominator? Angular momentum.

Time, like classical physics formulas is closed to new methods, new motives and even new means.

Some might remember how hard it was converting from "standard" units to "metric" units. Especially those factory workers and mechanics that had to transfigure both tools and fasteners just to do the same building and the same fixing of equipment. It might have been impossible for some.

Traditional classical physics have their proper place. But not multidimensional, multi-perspective, multiple inner revolutionaries perpetual with respect to systems, not time. In other words, open-open systems.

Since I was a child I, like many, could understand these yet unexplained dynamic forces. As many gymnasts, B.M.X.-ers and motocrossers can sense.

Throughout engineering school and beyond I never witnessed a solid reproducible explanation. Only recently could I investigate them with this new experienced physics. It has been a lifetime of work, understanding, and learning (much of it trial [literally] and error) that has led to today's venture and vision.

God's first of many Soundproof Safety projects has proven to amplify power, and sustain itself with fractional feedback. Making it an open system. And a new method to produce electricity, or energy. The seeds, I believe, to a revolutionary new day and age.

And it should prove to be the beginning of establishing His theories of I R Physics meant to become irrevocable Laws of science. Yes, God's I R Physics.

Looking at computer animated models of weather patterns today, or of an asteroid hitting the water on a computer generated animation, it's easy to see that the angular momentum aspects of I R Physics, that account for known centripetal forces, are not represented in detail. Probably because no current Laws of classical physics describes them entirely to represent them via computer programming. Yet, they must obviously exist for all matters that have angular momentum when revolved. Even if just gravitational "cannon-ball" like forces like those that whip spaceships out of our atmosphere and into orbit.

Particles obviously interact even if they do not touch. Therein lays the threshold (R.A.M.A.I.R.) to study.

If anyone has held a spinning bicycle wheel horizontal and tried to rotate it, it should be easy to understand by example the effect of centripetal geometric forces in nature about multiple, even variable axis. Yet they are currently very difficult to teach, learn, and understand within the established means of college books, universities, classical Laws, etc. "Maybe" they can't relate.

Relevant intensity, therefore, is a function of the dynamic leverage of angular momentum in open-closed systems. Because the faster the wheel spins, the more difficult the closed axis is to rotate.

And the relevant intensity is a function of dynamic kinetics in open-open systems. Because the more the duration and the degree that two weather systems meet, the worse a storm is. And the more potential for transformation and/or transfigured output.

If the dynamic leverage and the dynamic kinetics of angular moment were not a viable force, then I don't believe that freestyle motocrossers could do flips, that surfers could ride inside of water tunnels, that gymnasts could spin, or that pitchers could throw curve balls. Let alone there even be tornadoes, lightning, critical temperature for superconductors, and every other "unknown" force currently in debate. It is that transformational and/or transfigurational, I believe.

So, God blessed to me, and I to you, the brand new rock of I R Physics. The dynamic leverage aspects of revolutionaries, and the centrifugal geometrics and/or mass density of angular momentum systems that may answer many wonderful, applicable conundrums that science and society have yet to properly share. And thus, we haven't been able to practically use. Literally, a gift from Heaven.

And if these theories are proven Laws and applicable, then we may one day be able to shoot tornadoes away, more accurately predict weather patterns, predict multidimensional flow kinetics, clean up mother earth, have forever free electricity, a natural and a very prosperous, sustainable society and economy again. It can advance every true branch of education.

I R Physics with inherent R.A.M.A.I.R. Effect is that transfigurational (figuratively, literally and universally), I believe.

"The observer, when he seems to himself to be observing a stone, is really, if physics is to be believed, observing the effects of the stone upon himself."
- Bertrand Russell

9. A new look for the universal equation

$$F = ma \pm \, ?$$

Think outside the rotating sphere!

It seems self-evident to me that the main-stable of classical physics, $F = ma$ (force equals mass times acceleration), cannot itself be a universally complete equation because it does not take into consideration the open field affects that any closed force causes to an open system that it functions within.

$F = m$ (mass) times a (acceleration) may itself be subject to conditions. (Obviously.)

$F = ma$ only considers a closed, mutually exclusive force. (Yah...you might say, but … blah blah blah.)

For example, when a rock is dropped, $F = ma$ cannot take into consideration the dynamic kinetics of the atmosphere affected.

Therefore, $F = ma$ is a mutually exclusive formula.

However, a mutually inclusive formula (universal) must equal the closed system cause plus and/or minus the open system affects.

In other words, $F_{universal} = F_{closed}$ and/or mutually inclusive to F_{open}.

Where F_{open} = The relevant R.A.M.A.I.R.

F_{open} can have variable degrees of angular intensity with respect to F_{closed}.

For example, pressure can affect angular momentum. Temperature can affect angular momentum. Friction can affect angular momentum. As can shape, material and/or mass differences.

Therefore, R.A.M.A.I.R. must be a function relevant to these mutually inclusive variables.

And since F_{closed} is well known, the only variable to better decipher is F_{open}.

Which "should" yield a Universal equation?

A Universal equation would have to take into consideration the angular momentum affects in the open system, as well as the centrifugal geometrics and/or mass differentiation that effect the closed system angular momentum cause. But it's a small window.

The function of F_{open} must take into consideration the dynamic angular momentum caused by F_{closed}.

Mass differentiation within the closed system and open system may cause an additional dynamic leveraged, and/or dynamic resistive aspects to the mutually inclusive angular momentums.

Extreme temperature differences, or the speeds in which particles can travel, is an aspect that also affects mutual inclusiveness.

And the transformational degree needs to be a function of density, degrees, geometry and intensity with respect to angular momentum, irrelevant of the classical time scale.

R.A.M.A.I.R. may equal a percentage, or variable relationship of mutual inclusiveness. A true, not "false" scale.

However, we may never be able to put any permanent label on open-open systems. That seems self-evident to me. But these new concepts of I R Physics should encourage sharing understanding. Especially by and for those who experience it.

How does one relate open systems with closed variables or equations? Experience, obviously, plays the big, first role.

Let the static $F = ma$ now equal ma mutually inclusive to dynamic leverage, and let's see who can solve this equation first. And let's see how far we can take this.

In essence, there should be a variably dense sum of torque patterns that can also predict the relationship of R.A.M.A.I.R. to $F = ma$.

This expands the possibilities to pursue. The field of physics should now be wide open again.

"Keep in mind that if you take a tour through a hospital and look at every machine with an on and off switch that is brought into the service of diagnosing the human condition, that machine is based on principles of physics discovered by a physicist in a machine designed by an engineer."

- Neil deGrasse Tyson

Designs

10. Kinetically dynamic power technology

{A physical power amplification R.A.M.A.I.R. system.}

A. A duel axle, dynamic leveraged, amplified torque spinning wheel system for positive energy production by means of maximum torque applied at specific position of lower variable torque axis:

Picture a vertical free-spinning wheel on a fixed axis. (See next page.)

Now picture a second horizontal fixed axis. This second axis is horizontally-centered half way between the first axis and the wheel circumference. And mounted only on one side of the bigger wheel so that both can freely spin.

A weight is fixed with respect to axle two and able to freely slide along a radial arm that is a spoke of the bigger wheel on axle one. And a torque is applied at axle 2. A variable torque is induced at axle 2.

Top view of this duel axis torque amplification wheel:

mass

(The mass can rotate/slide,
but fixed to the end of arm 2.
The mass can freely slide
along radial arm 1.)

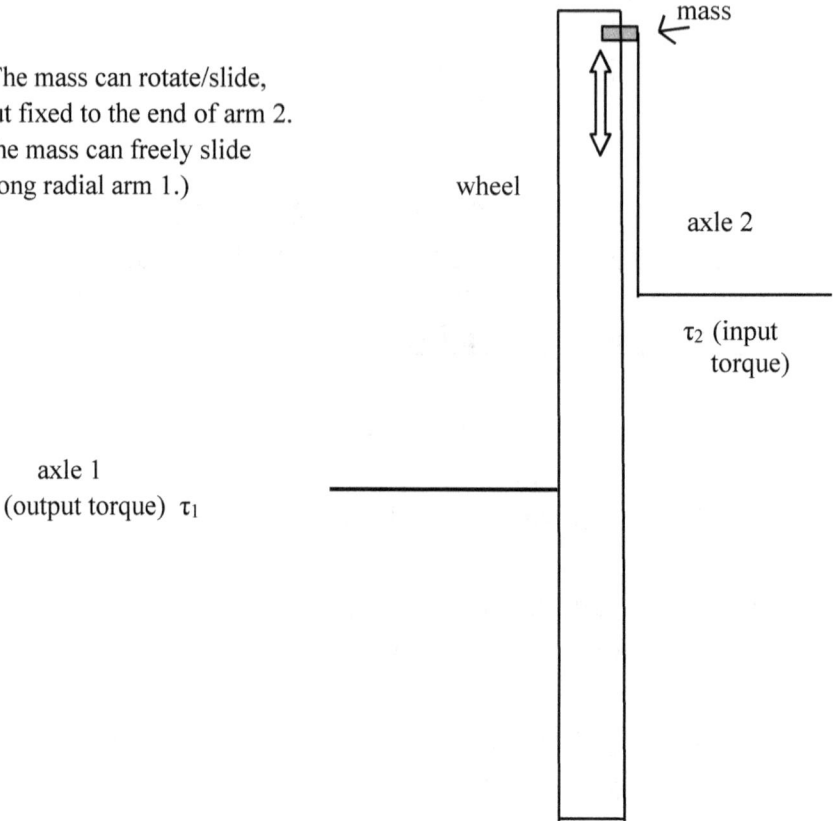

wheel

axle 2

τ_2 (input
torque)

axle 1
(output torque) τ_1

Side view of the duel-axis torque amplification wheel:

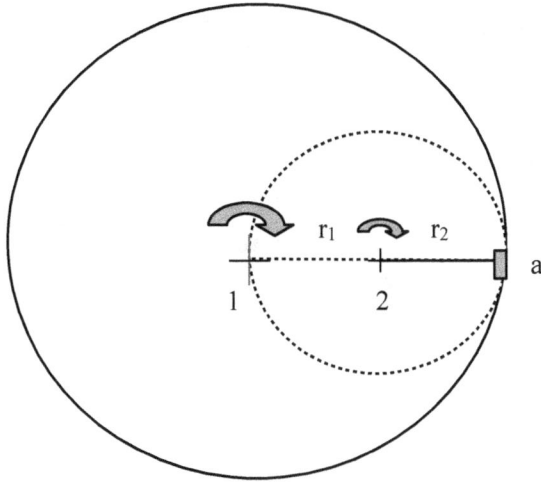

r_1 = radius of bigger circle
r_2 = radius of inner circle, equal to ½ r_1

Looking at the system from the top view, τ_1 (output torque) is produced relative to τ_2 (the input torque). Therefore, a dynamic leverage of angular moment is produced on the wheel at its center, or τ_1, by means of τ_2.

Looking at the system from the side view, after τ_2 is initially applied, the external wheel should produce a greater running torque, thus a positive net energy at τ_1.

Regarding torque: $\tau = r \times F$ or $r \times mv$ which equals angular momentum L.

When mass is at point a: τ_1 (output torque) $= \tau_2$ (input torque)

It seems intuitive to me that the net angular momentum and torque to the right of axle 1 can only be positive in the clockwise direction. And when the mass is near axle 1, there would be very little counterclockwise force acting on the net wheel torque and angular momentum because r approaches zero. Thus, a self-induced perpetual torque. Or, a variable positive power factor.

When mass is anywhere but point a: τ_1 (output torque) $= \tau_2$ (input torque) $\pm \tau_u$

(Where τ_u is the inherent ulterior torque.)

It's true that there would be some frictional limitations. However, when these studies are viewed in the electromagnetic realm, frictional limitations would be virtually negligent.

And it's true that the angular momentum of a duel axle system would be variable, but this could be com-pensated for by multiple axis and/or variable strength internal magnets of the receiving generator. Not to mention flexible axle arms and/or connection points.

It's the little differences themselves that may very well be the best & brightest future of all mankind.

It will not sound practical to some, but this is soundproof to others who have experienced it.

I believe that this is where the future of science, technology, economy and society resides.

No matter what the mass of the weight or the lengths of the radial arms (as long as radial arm 2 is less than half of radial arm 1) the dynamic leveraged torque, or mutually inclusive energy, looks to be self-perpetuating, maintained, easily controlled variable to the input energy. Hence, inner revolutionary physics of an I.E.D. net positive power system.

In a properly balanced system, the angular velocity, hence potential output energy, becomes infinitely controllable, variable and self-sustaining if electrical by an input energy that is much less than the output energy. A feedback control to produce net positive amplified torque, hence energy system.

As long as input power is less than output power, one owns perpetual energy potential.

If one horse is inside of a "hamster" type wheel, I believe that the created torque at the wheels center becomes a function of the radial arm length of the wheel.

If the radial arm length of a wheel is 10, isn't the single horse able to produce 10hp?

If the radial arm length is 1000, isn't the single horse able to produce 1000hp?

Start-up torque should require more energy. But self-sufficient torque looks to me to be very similar if r equals 10 or if r equals 1000, minus any friction.

B. A gravitational tri-axis, dynamic leveraged perpetual torque of a positive energy system:

Top view of the three axis torque amplification wheel:

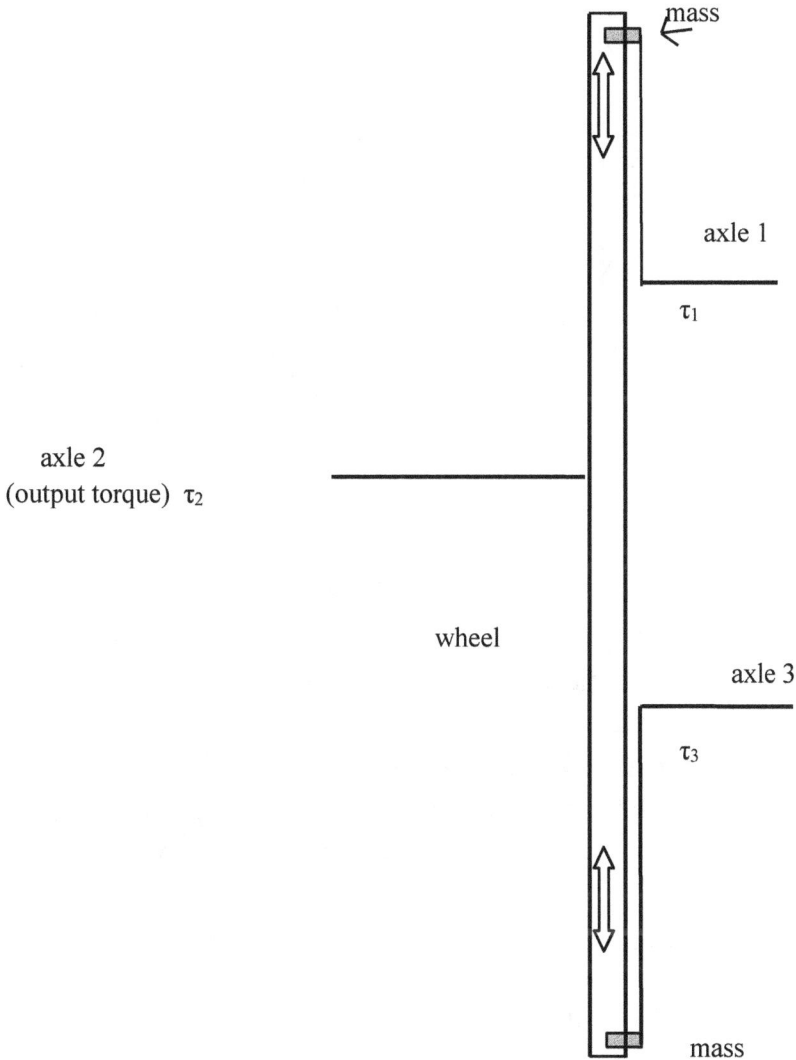

mass

axle 1

τ_1

axle 2
(output torque) τ_2

wheel

axle 3

τ_3

mass

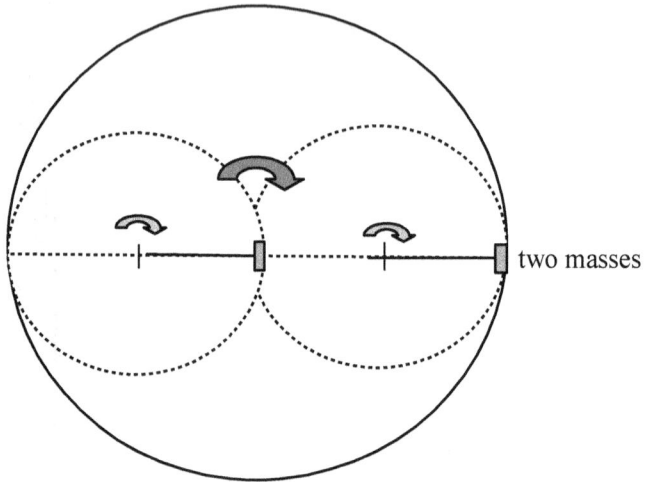

two masses

The output energy of the external wheel should be greater than the two torque inputs. In an electrical system, the input torques could be applied at specific times that would thus amplify the output torque and reduce the energy needed to provide the input torque as well (since it only needs to be applied at the top half of the rotation cycle). This would be a momentum dynamic leveraged system in gravity. And dynamic kinetics in an electrically actuated system.

If these two torque input radial arms leveraged with springs pulsed at the bottom dead center (for the left arm) and top dead center (of the right arm) then they could provide additional angular momentous forces for additional energy amplification. An additional positive dynamic kinetics potential. (Or variable positive power factor with respect to shared angular momentums.)

Ultimately, scientifically, if the exterior circle did not exist then the two smaller inner angular momentums could drive an ulterior shaft on their own.

C. Electromagnetic dynamic leveraged perpetuated torque and positive energy production method:

In an electromagnetic environment, or motor-generator concept, the gravitational mass could be replaced and/or added with a magnet or electromagnet, at the center of the system, and even at strategic locations outside of the system. These electromagnetic forces could actuate the system by minimal impulse currents, and produce a much higher output torque current.

Since torque is rotational energy, this system could be variable, especially with electromagnets and specific electrical impulse switches or programs. This system would be self-sustaining, and used for many applications given that centripetal force is all that's really needed to produce electricity.

And since work is relative to the radial arm length of the system ($W = r\mathbf{F}$), infinitely variable and controllable positive power production should be possible. Even with minimal feedback for the input. Rendering the system consumption and emission free of pollution resources.

{i.e. ∞ *energy with no consumption and no pollution.*}

Theoretically, one could pulse a magnetic piston with respect to an electromagnetic spark plug in motor cylinders.

I believe that dynamic leveraged centrifugal torque and/or even spherical mass torque, could easily amplify electricity, and reduce carcinogenic energy

110

consumption. And adding magnetic torques can more efficiently amplify the energy output. Low voltage into a dynamic leveraged system to produce higher voltage output. The system could be self-perpetual by fractional feedback current. A brand new circle of life, so to speak.

I believe that a bicycle wheel with the forks attached hanging upside down could be made to fly with dynamic kinetics inner revolutionary physics. Even with the use of gyroscopic directional remote control. (This is just an example of the concept in application.) I know that it's all pretty hard to comprehend initially. But it's easier to understand if you experience it.

In a dynamic leveraged system that produces centripetal force, the initial maximum force velocity is all that is required to maintain a net positive torque because of angular momentum. The output produced, therefore, is greater than the input needed to maintain a perpetual motion that could generate electricity.

In my BMX and MX days, I always wondered how some riders could get better lap times simply by the way that they navigated jumps. Although classical physics has yet to explain their secrets, I believe that the dynamic leverage of angular momentum method is getting close.

In riding a bike there are at least two centripetal force factors at work (even more if we consider front and rear wheel spin or motor RPM variables.)

The bike and the human body are the two main forces. When a rigid body on a bike goes over a jump, that distance can be calculated and pre-determined with classical dynamics formulas.

However, when a body absorbs the upward and outward momentum of the jump, that distance is reduced. And those are the faster riders. And that's an example of the benefits of experiencing the dynamic leverage of angular momentum.

And two or more dimensional dynamic leveraged suspension concepts would increase the stability and increase lap times as well. Because it adds another angular momentum aspect to defy obstacles.

On today's mountain bikes and motorcycles, the one dimensional suspension absorbs vertical impacts. If a suspension system was added horizontally to the swing-arm and/or forks, then this would absorb the horizontal components at the encountered with respect to horizontal speed.

And the combination of these two would absorb a more accurate angular momentum suspension aspect needed.

A gymnast on uneven bars would be another example of the dynamic leverage potential of angular moment. This system obviously defies classical conservation of energy theories.

The best gymnasts on uneven bars understand the dynamic leverage of angular momentum. They know how to gain power by flipping their body with respect to semi-flexible uneven bars. There arms and/or flexing bar acts as usable suspension and/or spring assistance

In classical physics, "conservation of energy" says that one rider cannot be faster than others just by moving his body properly over obstacles with respect to a bike in order to take advantage of angular momentum.

But any professional rider could tell them that isn't true. And I've verified this many times on the track myself.

Proper angular momentum is a most beneficial source of net positive power production and inclusive energy studies.

Dynamic leverage of angular momentum that itself produces positive net torque is a force yet to be well studied.

With the study of I R Physics one may very well be able to determine when and if an unbalanced tire on an automobile would blow out.

Likewise, when and if an unbalanced earth rotation may get more unstable as it has recently by the shift in axis and/or mass density variables.

Once these factors of layered centripetal systems can be properly formulated for broad application we'll have our answers to catalyst free, pollution free, clean energy systems on any and every scale needed, I believe.

We may escape pending doom that we don't even currently know exists.

And we'll have our answer to cleaning up the atmosphere, mother earth and much more once these answers are mass produced and then applied into action.

11. A new method for generating electricity

A new method for generating electricity for application.
The positive dynamic leverage of angular momentum.
The positive power factor affect.

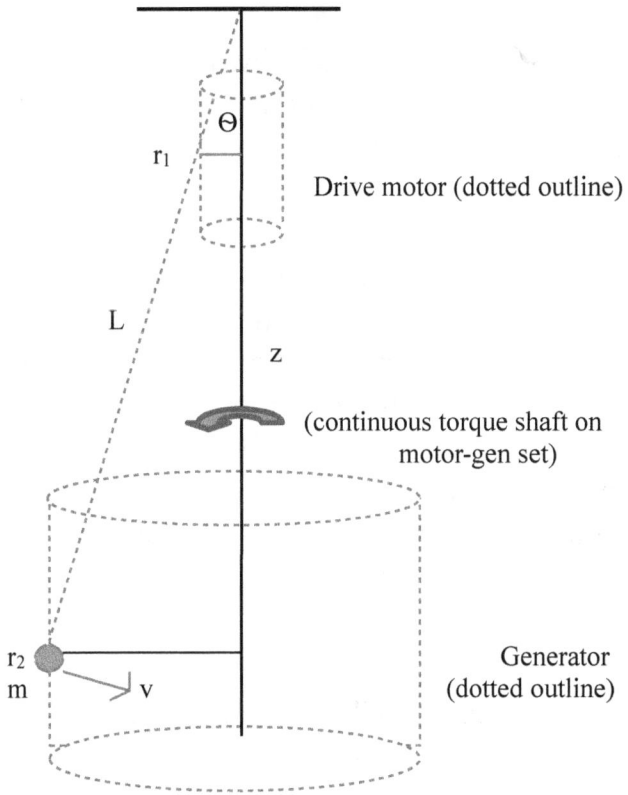

Θ

r_1

Drive motor (dotted outline)

L

z

(continuous torque shaft on
motor-gen set)

r_2
m

v

Generator
(dotted outline)

From Newton's 2nd Law: $\sum F = ma = mv^2/r$. Solving for time anywhere along L:

It's known that t (rotational period) = $2\pi(Lcos\Theta)^{1/2}/g$

Therefore, it is known that the period does not depend on m (the mass of the particle). This mutual exclusiveness is what we're trying to exclude via I R Physics.

For any and all r's (variable radius') evaluated, τ (torque) is constant up and down the vertical shaft z. And the net force becomes a function of rotational speed and/or mass differentiation.

It is also traditionally believed that potential energy is a characteristic of a system, and not the individual components within the system.

But that's what we're trying to help.

Positive power torque theory within an electromagnetic "conical" centrifugal system:

Since torque (τ) = r x mv, τ is proportional to r throughout z.

And the ratio of r_2/r_1 becomes the estimated positive power factor for this dynamic leveraged angular

116

momentum system with a common torque motor-generator set.

<u>In electromagnetic torque (the motor-gen set):</u>

$\tau = r \times \mathbf{F} = \mu \times \mathbf{B}$, where $\mu = NIA \times \mathbf{B}$

Were μ = the magnetic dipole moment, N = # of turns, I = current, A = cross-sectional area, and **B** is the magnetic flux.

For two devices with coupled torque:

$\tau = N\,I_1\,A \times \mathbf{B}$ (for system 1) $= N\,I_2\,A \times \mathbf{B}$ (of system 2)

It seems counter intuitive. But I believe that if Area 2 equals twice area 1, $I_1 = \frac{1}{2}\,I_2$. And equally proportional to N and **B** for both systems. This means that output current is always greater the input current for this particular example.

In other words, the electricity out is greater than the electricity in.

<u>If this theory becomes a new law of physics:</u>

Since work equals torque, work would be a function of r_2/r_1 with respect to Δm (change in mass) times the velocity which could equal a Positive Power Factor (PPF)

117

In theory, the work at r_1 could equal 0, and the work at r_∞ could equal ∞.

Additionally, the physical model of this dynamic leveraged centripetal conical system could add additional torque to wind turbines by simply reducing the angles relative to the torque shaft in application.

Looking at author's first free electricity project with respect to a virtual conical pendulum:

In my first experiment to prototype this concept, I directly coupled together a 12 volt D.C., 0.6 amp drive motor to a generator (12 volt, 70 amp capacity) to produce 5.2 amps at 12 volts D.C. output. That's a positive power factor (P.P.F.) of over 5 produced within a single system. The R.P.M.'s were not matched.

By matching R.P.M.'s that produce the most efficient electrical generation and drive, I see no problem increasing this P.P.F. to 10 or more. While at the same time running the drive motor with fractional feedback power from the generator.

With more test confirmations, this is a true perpetual (which I call power amplification) system that self-sustains itself after start-up to produce net positive electricity within itself.

Direct current is more efficient than A.C. because there's no impedance or resistance due to oscillation waves of the electricity. So an A.C. inverter can be added to this system to a current sinking capacitor and/or battery.

A battery is needed for D.C. applications for start-up, storage and capacitance.

Since v (rpm) can be made known and variable, average r's of driver and generator are known and variable, and torque is common, a positive power factor by means of a negative feedback current should be produce-able and mass-producible. And when it is, it can be applied to any size or application. Without the need of transmission lines or polluting fuels. It can also be built into a single housing. It can be portable or fixed depending on one's need.

Although "conservation of energy" in classical physics says that this system is impossible, I have proven that it works.
When "conservation of energy" develops to consider centrifugal geometric systems, a potential unlimited number of shapes can be studied. Including, hopefully, a simple rod shaped motor-gen system.

In every yet unexplainable aspect of physics (which explains almost every other productive field of knowledge) one can know intuitively that there are dynamic leverage aspects at work.

Wemple's theory of perpetual R.A.M.A.I.R.

Forces of universal attraction exists for com-petitively complimentary, mutually inclusive systems where there is a balance in a duration within an intensity that can and will produce amplified power.

In other words transfiguration exists.

For example, when R.A.M.A.I.R is greater than the Transformational Degree (T.D.) of the duration caused by the maximum only I.E.D. velocity of an internal wheel reacting within an external wheel, but stays below the Transfiguration Zone (T.Z.) I.E.D. intensity, a net E.I.D. positive power factor greater than one is produced and sustained. (For open-closed systems.)

An increasing duration plus an increasing intensity equals the systems perpetually positive T.D.

An increasing duration plus an increasing intensity plus a competitively complimentary, mutually inclusive location (for open-open systems) can equal perpetually positive T.Z. (Transfiguration Zone) momentum.

In other words, something out of a previous nothing.

Dynamic leverages of centrifugal geometrics and/or mass variations are forces yet to be well studied. But are now better defined as they are known to exist.

These forces are our key to a revolutionary new day and age.

For I R Physics is a new rock to advance every other true field of science, study, technology and application.

That's my theory anyway.

12. Light spectrum waves

I suspect that various waves may metabolize (absorb) at different rates for different people like different material makeup. Therefore, wave speed may not always be relevant to visible light. Absorption rates (like on/off speeds) may be what's humanly relevant for each person's material make-up and absorption rate (even for visible light absorption in our eyes). Anyway, it's another dimension to consider.

Nuclear radiation may be isolated and absorbed someday. "Autism" and other mutually exclusive systems may be better understood and treated as a virtue of different perspectives of different senses. Finding and building off of the mutually inclusive aspects are what may be relevant.

A multiple perspectives study of these issues can only help.

Cancer research may someday have a new approach. As far as the best dynamic leverage and/or leverage densities to use in order to dissolve the matter and let other parts of the body absorb it to purge it out of the system.

I believe that energy can now be better investigated from multiple dynamic leveraged perspectives. And the R.A.M.A.I.R. transfer motive from various perspectives may be a hidden virtue waiting to be investigated.

The Laws of Thermodynamics are already there.

I believe that time is only a symbolic tool. I believe that the space around us is constant, but composition and density are variable.

I believe that light could have inner revolutionary qualities to consider. And not so much time value constraints.

I believe that if light were constant, then it could not be deflected in magnification and a mirror. I believe that there are inner relations between light density and magnetics, not light and time.

I believe that light should be studied in multiple ways; viewed from the source, reflected off material from the source, and that accepted into the eye/brain system. And all should consider the absorption rates of the instruments used (including human eyes, front to rear deflection to focal point), etc.

When we "see," it may be just a matter of energy (dynamic leverage) transfer and absorption rates, physical abilities, rather than something constant. In other words, light may be R.A.M.A.I.R., but not I.E.D. or E.I.D.

I believe that a butterfly around the other-side of the world can't cause a hurricane here, but 3-D dynamic leverages of atmospheric concussions (accumulated R.A.M.A.I.R. dynamic effects) from a small catalytic source can.

I believe that negative impressions may be absorbed into the mind like many other frequencies, if the host accepts it, and has enough contact time.

I believe that positive impressions are much more contagious given proper duration of acceptance/agreement. Being that the positives are more productive, those should be postured to suit their glory and efforts

for the most probable exposure, acceptance and perpetuation. A positive only, holy voice can only intensify
this exposure to perfection.

"State" "law," statutes, court, and life are
condoned and enforced from only a judge's perspective.
It could only benefit we the people if it were
upheld from our perspective. We the peoples
perspective with respect to a Holy Constitution.
Never should we just assume some are right
because of their title or marketing tactics. That
eventually has to become delusioning above all.
Trust, but verify. And only then, accept and obey.
(But that's just my own personal soap box speech
anyway.)

13. Degrees of I R

The degrees of true (inclusive) and the "degrees" of false (exclusive) scales are most likely the best method for advancing the study of I R Physics with respect to dynamic leverage in open-open systems. But the Degree scale in I R Physics only measures the true (mutually inclusive) aspects.

Therefore, I R Physics is the study of the mutually inclusive aspects, or relevant aspects, to systems with respect to angular momentum (I.E.D., E.I.D. and/or R.A.MA.I.R.)

If one where blowing bubbles from a soapy water mix and two bubbles become one, then the transfiguring R.A.M.A.I.R. could be understood as the I.E.D. (the smaller bubble) plus the E.I.D. (the bigger bubble.)

But if the two bubbles only partially stuck together, then the relevant R.A.M.A.I.R. degrees transformed would become proportional to each side, but different for each side, and produce an ulterior angular momentum. This can be tested.

And if three or more bubbles stuck together, then this I R method could evaluate the system as a whole based on the degrees required to produce ulterior angular momentums.

I R Physics, therefore, is the study of the mutual inclusive aspects of dynamic kinetics. Or, of motions changing while in motion.

I R Physics is a study of the degree of mutually inclusiveness relevant to the proportional causes of ulterior effects. As well as the transformation degrees required in order to possibly produce transfiguration.

It may never be stable for systems in transition. But transfiguration may available where it's positionally known.

There are degrees where open-open systems are transformational. And there is a duration of intense degrees plus location in which open-open systems birth transfiguration.

I call this T.D. (Transformation Degree) plus location potential the Transfiguration Zone (or T.Z.)

Closed systems are subject to open systems by nature.

The bottom line is, every new truth came about as I R. And every known truth can then be reviewed, better understood and better known by continuing this means of reference and perpetuation.

I R Physics can be thought of as classical physics, but physics amplified. A win-all for advanced education and/or learning and/or application.

A continuous relevance in transformation, figuratively, educationally, economically, socially, and literally.

I R Pioneers

When things get too complicated and/or too established we seem to step all over ourselves instead of progressing on ahead for ourselves.

I'd like to see new visionary and missionary team members interested in helping different, but complimentary ventures like the Upright U.S.A. Race Family Matters Mission alongside I R Physics. Means and efforts in these ventures, Race and mission looks to be minimal, but perpetually prosperous.

Holy members do not promote themselves until those underneath of them have passed elevated. For that's God-ism. The epitome of this is Jesus Christ.

S.S.R.T. (Standard Symbolic Represented Truths) would be a new (even if improvised or virtual to start) and much needed research and study to begin improving this newly discovered field of physics.

Our initial stage is to simplify and then more efficiently standardize symbolic classical physics so that it may be better taught and understood. Upright U.S.A. holds all Rights, Interpretations and/or Responsibilities and will be the directing manager to compliment business, education, and society.

God impressed upon me, "I R Physics." Its discovery has the potential to redevelop the world as we know it. From the first of many innovations, "Forever Free Electricity," through current studies, "dark matter" (which I call "enlighten science"), to several other advanced clean energy technologies, its potential should be self-evident and self-sustaining.

A high caliber of vision and mind power in team members with experienced understanding is needed here. A new S.S.R.T. also keeps these new discoveries and technologies fresh and competitive longer.

Advanced animated computer modeling, computer programming, web development, publications, and media production team members are needed as well.

Team Sound Proof in life science and the science of self-evidence aims to inner revolutionize stigmatized society groups. Constitutionally squared being the principle here.

Much manipulation, financial "legal" redundancies and never self-checking has put many States, citizens, and families in the red. Financially, psychologically and physically. Economically and, thus, socially.

Common sense could guide these systems better than the course they (we) are on now. Constitutional Law and State team members may be needed here.

The Family Matters Mission of America will be a benefactor of our pending parent organization. Its need should be self-evident and understood better considering my class action complaints in Federal Court, Central Illinois Division, Springfield, IL, of the current no Due Process for the old "family" court. It's need is known by all who have suffered from inherent consequences of the current establishment of "family" courts.

The Upright U.S.A. Race will be seeking drivers for patents and publications over and/or under Federal laws and/or leverage "laws" to defy governmental complacency.

And if need be, all rights of all new products and

technologies will be waved for an independent Upright U.S.A. (A little leverage for reform and/or our own patent office to compete with the current out dated and/or sabotaged and easily exploited patent office.)

I believe that "physics" are not outside of God's realm. Physics are simply symbolic truths used to teach and then apply what already is. Truth is truth.

Without Jesus, none of this would be possible. He gave me I R Physics to capture and re-develop what others have used and abused. Therefore, He gets all of the glory.

I believe that State Courts are not justice. State Courts are only condemnation pits. They are the idol of enmity upheld inherently shoring divisions via Weegie Wedging "magic" when we're supposed to be inherently uniting, not inherently dividing.

It's perpetually negative side effects on our children and families have left our great nation in a depression of oppression by & for misguidance. It's unintentional, just the nature of two opposing positions within a known broken system. But one that can "legally" gain control itself by & for any means necessary.

I've discerned that people cannot be law them-selves. And a new democratic discernment method, motive and means can be the only truth in justice and in life. And a proper constitutional judicial system as well.

Marketing looks to be self-evident and self-sustaining.

Competing with all the current status quotes looks to be the best venture for a new and much improved revelation. The Sound Proof Upright U.S.A. Family Matters revolution.

Even a recycled State system of solutions with help from the Upright U.S.A. mutually inclusive formula.

A competitively complimentary safe, sane and most civilizing balanced approach.

And it will be much needed in the days ahead for a balanced, national stability whose laws and procedures can only get more and more unstable. Thus, more and more oppressing to We the People.

Together, I believe that great things lay on the horizon ahead.

But, I can't do it alone.

And even another tomorrow is a blessed gift from God.

Applications

Social implications and/or applications for practice

14. The transfer and/or transfiguration to positive power dynamics

Positive growth energy transfer dynamics began based on classical thermodynamics study, but now with an added R.A.M.A.I.R. element that can prosper systems rather than decay them. A positive growth "pheno-menon" that can be incorporated into every other known field of science, biology, psychology, society, technology, etc.

Biblical scholars already know of its divine superiority. So I'll go ahead and try to explain it.

Physical life could not live for very long if the suns fire burned out tomorrow.

Likewise, the intellectual human psyche could not live long if God is left out. And lately there is so much other noise out there to distract us that our fire that it is getting dim.

The Laws of classical thermodynamics prove that all non-living systems must only decay over time.

Without an outside competitively complimentary force human psyches are no different. For without positive outside guidance, something to aspire to, it's too easy to get lost in the noise and/or another dying thing.

Therefore, there is no possible way that this could have ever been possible, or maintained without a living Jesus Christ and/or Christians today. For at the time, He was the only one to stand against systems of inherent social decay.

And there's no way for Him to have known the benefits of this unless by God.

The good sportsmanship concept of competitively complimentary relevance to inclusive methods, motives and means are the only good opportunity, the only viable option left when sin goes too far. But it's a formula for success, none the less.

Because it's a formula of attraction, not division. And that equals you squared. Therefore, the analogy $u^2 = c^2 \| m^3$ is the analogy that I use to remember this success formula.

When sin is superior to society in the eyes of our govern-mental beholders, then things have gone too far.

When you're left to inherently depend on sin then you are helpless and hopeless.

But faith, love, truth, mercy and forgiveness are still viable options for the asking.

Because either way those attractive attributes are known to win out over time, duration and/or intensity.

Therefore, favor relevance, not pestilence.

15. The Family Matters Mission positive enlightening dynamics
{The Upright USA Race Family Matters Mission}

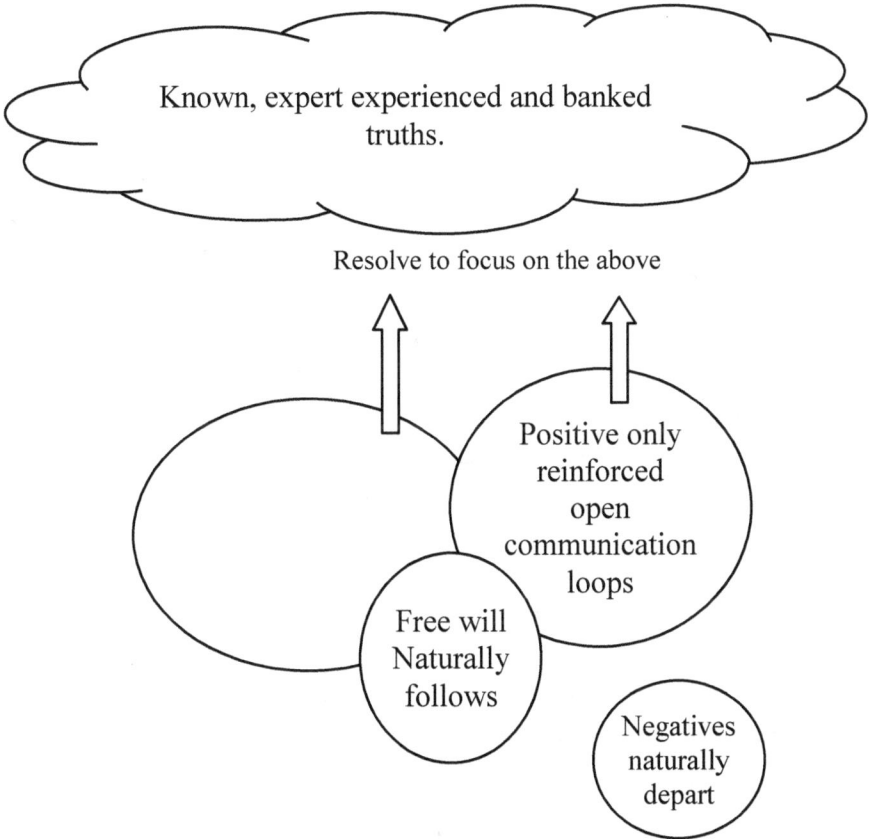

Known, expert experienced and banked truths.

Resolve to focus on the above

Positive only reinforced open communication loops

Free will Naturally follows

Negatives naturally depart

A mutually inclusive overlap of the three or more member guidance system should be the uniting points. The best method is positive communication and believing. The most secure is accountable physical authorizations.

Less negative plus more positive plus time equals more positive growth. Of course balance is the key here. A little with truth is more than infinity with no truth.

There is the strongest potential if and when the strongest guidance is correct. This is easily understood. But highly opposed as it inherently discredits job titles.

Love & Truth

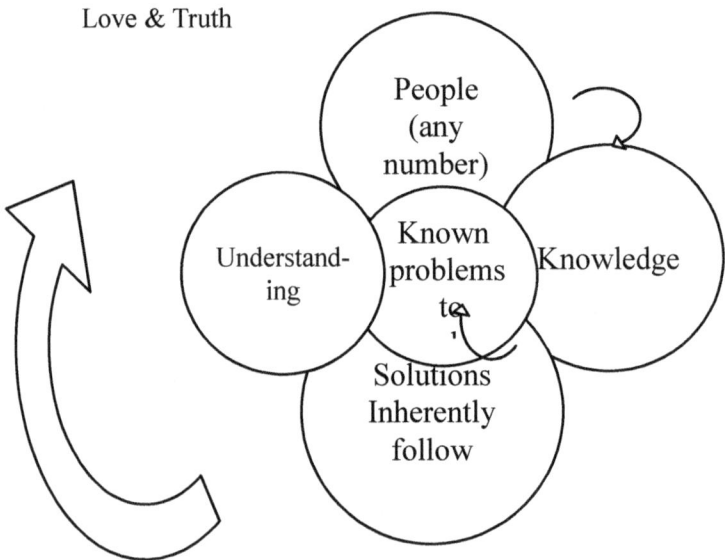

People (any number)

Understand-ing

Known problems to

Knowledge

Solutions Inherently follow

This second group handles the negative aspects of group dynamics for problems to resolve. It's more social, but less personal family oriented.

Gossip and false witnessing hinders progress. Therefore, all-inclusive communication is vital inside of this group itself. You can't fight water with no water. Above all we have to forgive. And that goes both ways. One cannot exist for long asking by neglect to be stabbed in the back.

This whole system of Team Sound Proof naturally drifts onward and upward over time.

Therefore, these two family systems attract each other to even build each other up.

It's the natural and now known science of obvious self-evidence relevant to any and every problem and/or group. Mutual inclusiveness is the key.

Though we're known to misplace our keys.

For each group/family/judiciary/government/etc., balanced positive growth equals self plus a flexible positive model discernment method plus communication. (Or R.A.M.A.I.R.) And this proper medium equals an overall balanced positive power factor back to people, by people and for the People.

This Upright USA Race incorporates:

- A focused development on the mutually inclusive aspects;
- Apply the $c^2 \| m^3$ formula as a game;
- The most dedicated groups already exist;
- Disciplined guidance subject to students;
- Loving those opposed to truth even more;
- A Resolve to own truth in groups (never singular);
- Do know by understanding;
- Serve and never assume;
- Only legitimate experience can be legitimate witnessing;
- Help each other understand rather than "helping" each other hurt.

Above all, enjoy the process and be inspired by & for helping change the world. Together. Forever.

16. The Upright U.S.A. Race

Start seeing new stars!?

We'll have to see about that. Obviously a lot of inspiration and/or work needs to be done.

But the absolute best work load is the right light load.

Test the read, know the light, and then be the switch. And enjoy watching others automatically turn on.

At the Sound Proof Safety Firm, all kinds of new revelations are being discovered. Discovered as a direct result of I R Physics.

With plans to bring together team members who have been back-bitten by political, professional, and/or govern-mental groups to pursue the challenges that we all face together head on.

Experienced pros, mentally and spiritually strong. Efficient groups, with enlightened justice, true love and worthy honor. Absolutely We the People, by We the People, in the most contemporary spirit of life to permeate into society a contagious joy of living, laughing, learning and advancing.

Weather by implosion or explosion, many societies let friction advance until they dissolve themselves. They get stuck and wait until it's too late.

Let compulsive inherent dividing no longer be us.

In life, there is no such thing as infinity. But there is intellectual infinity. And there is team building thanks to our most noble heroes.

I don't see the United States of America waiting until it's too late to stop inherently dividing and to start inherently uniting and healing.

I see us as a proactive group of pro people regardless of where our compulsions try to drive us in vein.

A people who are not afraid to call sin for what it is.

A group of people who are not afraid to prove that sin is worse than uselessness.

A group of people who cannot be converted contrastingly by stagnant status quotes.

I believe that our minds encompasses what we've accepted via our senses.

I believe that our spirit is what we project.

Delusional marketing has gone a long way to promote the ills of the world to be used against us.

But the American spirit is second to none, save one.

Good or bad at times, our hearts are built by these perpetual cycles and intensity to positive durations. A degrees of intensity that can only succeed.

But the prestigious pride in the dynasties of sins have gone too far.

Love united in action is the world's most

144

strongest force.

I believe that the love (or lack thereof) that we receive and realize to build upon from a child is the foundational emotion that drives a lifetime of mind, body, spirit, and heart power. And all together encompasses our human psyche.

Judicial government has overstepped its boundaries by dictating normal family dynamics in courts with unconstitutional sin. Thus, cursing the American spirit, the Christian spirit, and free will as a whole.

But that's not the worst of it. Complacent officials have aided and a bedded those suppressive and/or traumatizing wrong forces.

Though down, we're not out just yet. We can all be heroes by being an inspiration for a solution. (Known solutions of course.)

We should have free lives, free power and free help. Anytime, anywhere, and indefinitely.

In reality, we already do. It just isn't that convenient yet. Only Love can make it that much more convenient.

To sum it all up, We the People need the power and application of the Lord's I R Physics in a new and/or renewed university, in a new and/or renewed government, in a new and/or renewed judiciary, in a new and/or renewed legislation, in a new and/or renewed society, in a renewed community, in new and/or renewed energy sources, in a new and/or renewed mind sets, and in our own lives today anew.

And standing beside the known mind sets of truth, prosperity, love, honor, family, faith, friends, and respect can only get us there anew.

Here's to a leaner, cleaner, meaner, more abundantly prosperous, more civiling, more sane, and properly free-er human existence.

Let's break those complacent chains that bind us to take advantage of us today, and take this world to the next level of greatness together!

Because only together, we can.

Never lose love, faith, purpose, mission, vision, or drive!

"The more that you seek, the more you will find. That's heavenly I R Physics. A river of life."

Albert Einstein once said, "Give me a long enough lever (enough leverage), and I could move the earth."

Well, I say, "Give me the right quality of internal dynamic leverage love, and we can all change the external world together, forever, for the better."

And Jesus may say, "finally, amen."

BELIEVE JESUS.

Acknowledgments:

Relief of burdens courtesy of Jesus Christ.

Cover text courtesy of Cooltext.com

Back cover picture – Visions by Carol

Photo edits by ipiccy.com

Quotes courtesy of Brainyquotes.com

Other books by author Aaron W. Wemple:

S H U S H E D *Supervoid Recovery Steps*
The Law Doctor,

The Upright USA Race

Love's Loop

with Family Matters Mission

www.ingramcontent.com/pod-product-compliance
Lightning Source LLC
Chambersburg PA
CBHW071858200326
41519CB00016B/4445